Chiheb Dahmani

Static & Dynamic Magnetic Fields in Nanoparticle-based Drug Delivery

Chiheb Dahmani

Static & Dynamic Magnetic Fields in Nanoparticle-based Drug Delivery

Südwestdeutscher Verlag für Hochschulschriften

Impressum / Imprint
Bibliografische Information der Deutschen Nationalbibliothek: Die Deutsche Nationalbibliothek verzeichnet diese Publikation in der Deutschen Nationalbibliografie; detaillierte bibliografische Daten sind im Internet über http://dnb.d-nb.de abrufbar.
Alle in diesem Buch genannten Marken und Produktnamen unterliegen warenzeichen-, marken- oder patentrechtlichem Schutz bzw. sind Warenzeichen oder eingetragene Warenzeichen der jeweiligen Inhaber. Die Wiedergabe von Marken, Produktnamen, Gebrauchsnamen, Handelsnamen, Warenbezeichnungen u.s.w. in diesem Werk berechtigt auch ohne besondere Kennzeichnung nicht zu der Annahme, dass solche Namen im Sinne der Warenzeichen- und Markenschutzgesetzgebung als frei zu betrachten wären und daher von jedermann benutzt werden dürften.

Bibliographic information published by the Deutsche Nationalbibliothek: The Deutsche Nationalbibliothek lists this publication in the Deutsche Nationalbibliografie; detailed bibliographic data are available in the Internet at http://dnb.d-nb.de.
Any brand names and product names mentioned in this book are subject to trademark, brand or patent protection and are trademarks or registered trademarks of their respective holders. The use of brand names, product names, common names, trade names, product descriptions etc. even without a particular marking in this work is in no way to be construed to mean that such names may be regarded as unrestricted in respect of trademark and brand protection legislation and could thus be used by anyone.

Coverbild / Cover image: www.ingimage.com

Verlag / Publisher:
Südwestdeutscher Verlag für Hochschulschriften
ist ein Imprint der / is a trademark of
OmniScriptum GmbH & Co. KG
Heinrich-Böcking-Str. 6-8, 66121 Saarbrücken, Deutschland / Germany
Email: info@svh-verlag.de

Herstellung: siehe letzte Seite /
Printed at: see last page
ISBN: 978-3-8381-3936-4

Zugl. / Approved by: München, Universität der Bundeswehr München, Dissertation, 2014

Copyright © 2014 OmniScriptum GmbH & Co. KG
Alle Rechte vorbehalten. / All rights reserved. Saarbrücken 2014

For my parents,
for my wife,
for my son,

List of publications originated in this work:

- I. Hoke, C. Dahmani, T. Weyh. Design of a High Field Gradient Electromagnet for Magnetic Drug Delivery to a Mouse Brain. Proceedings of the COMSOL Conference 2008, 4. - 6. November 2008, Hannover. (as paper presentation)

- Dahmani Ch., Götz S., Weyh T., Renner R., Rosenecker M., Rudolph C. Breath Synchronous Magnetic Drug Targeting in the Lungs. Proceedings of the 4th European Conference of the International Federation for Medical and Biological Engineering. Antwerp, Belgium, 23. - 27. November 2008. Volume 22, ISSN 1680-0737 (as paper presentation)

→ Obtained Award for Best Presentation at the DGBMT 42nd Conference in Antwerp, Belgium

- Ch. Dahmani, Th. Weyh, H.-G. Herzog. "A simplified Approach for Nanoscale Magnetic Moment Measurement and a Study of the Impact of Nanoparticle Interaction on their total Magnetic Moment". Proceedings of the conference "Seeing at the Nanoscale VII - Exploring the future of Nanotechnology Using SPM and related Techniques". July 28-31, 2009, University of California, Santa Barbara, USA. (as poster)

- Dahmani Ch., Götz S., Weyh Th., Renner R., Rosenecker M., Rudolph C. "Respiration triggered Magnetic Drug Targeting in the Lungs". Proceedings of the "31st Annual International Conference of the IEEE Engineering in Medicine and Biology Society", September 2-6, 2009, Minneapolis, USA. (as paper presentation)

- Dahmani Ch., Helling Fl., Weyh Th., Plank Ch. "An Innovative Rotational Magnetic System to enhance Cell Transfection with Magnetic Nanoparticles". Proceedings of the "World Congress for Medical Physics and biomedical Engineering 2009", September 8-11, 2009, Munich, Germany. (as paper presentation)

- Stefan M. Götz, Chiheb Dahmani, Carsten Rudolph, and Thomas Weyh. „First Theoretic Analysis of Magnetic Drug Targeting in the Lung". IEEE Transactions on Biomedical Engineering, Vol. 57, No. 9, September 2010, 2115-2121. (journal paper)

- Hanna Mannell, Joachim Pircher, Franziska Fochler, Yvonn Stampnik, Thomas Räthel, Bernhard Gleich, Christian Plank, Olga Mykhaylyk, Chiheb Dahmani, Markus Wörnle, Andrea Ribeiro, Ulrich Pohl, Florian Krötz, "Site directed vascular gene delivery in vivo by ultrasonic destruction of magnetic nanoparticle coated microbubbles"
Nanomedicine Journal, Received 30 August 2011; accepted 24 March 2012. published online 04 April 2012.

- Dahmani Ch., Mykhaylyk Ol., Helling Fl., Götz St., Weyh Th., Herzog H.-G., Plank Ch. "Rotational magnetic pulses enhance the magnetofection efficiency in vitro in adherent and suspension cells". Journal of Magnetism and Magnetic Materials, Volume 332, April 2013, Pages 163–171.

Book contributions:

Article "Design of a High Field Gradient Electromagnet for Magnetic Drug Delivery to a Mouse Brain" included in the book "Introductory Biophysics: Perspectives on the Living State" by Claycombe, James R. / Tran, Jonathan Quoc P., Jones & Bartlett Publ., Publication Date: April 1, 2010, (ISBN-13: 978-0763779986)

Acknowledgements

This book is based on the research activity I conducted at the Technical University of Munich between 2007 and 2010, and with which I obtained my PhD degree.

So, first, I would like to thank my parents and my wife for their continual love and encouragement through the entire span of my studies, my PhD research and beyond. Without their constant support, this adventure would not have been the same. They truly deserve a standing ovation.

I would also like to thank my advisor, Prof. Dr.-Ing. Thomas Weyh, whose dedication, invaluable guidance, support and ever-present optimism has fueled my ability to conduct this work. I am truly blessed and honored to have been mentored by him.

Special thanks go to Prof. Dr.-Ing. Hans-Georg Herzog for his help and assistance, and for the absolutely enjoyable ambience at his Institute for Energy Conversion Technology (EWT) of the Technical University of Munich.

Besides my advisors, my friend and colleague Dr.-Ing. Stefan Götz deserves a special mention. He always gave insightful comments to my work and bolstered me with a lot of fun, even in hard times.

I am equally grateful to my colleagues of the EWT workshop, Mr. Tuschl and Mr. Wild, for their help with the construction and implementation of the several technical concepts generated in this work, and particularly for their patience. Also thanks to the folks at the EWT for the funny time and the very interesting discussions.

I also would like to address many thanks to my collaborators at the Klinikum rechts der Isar, especially Dr. Olga Mykhaylyk, and Dipl.-Molecular Med. Katharina Schilberg, for their help with cell and in vivo experiments and for their valuable advice regarding biological and medical matters.

I finally would like to thank the members of the jury, PD Dr. Rainer Burgkart and Prof. Dr.-Ing. Rainer Marquardt for accepting to take on the second correction of this dissertation and the examination.

This research was supported by the German Federal Ministry of Education and Research (BMBF) under grant 13 N 9184.

Table of Contents

Abstract 3

Zusammenfassung 5

1 Background and Motivation 7

    1.1 Magnetic nanoparticles in medicine 10
        1.1.1 Manufacturing: 11
        1.1.2 Biomedical applications of Magnetic Nanoparticles: 14

    1.2 Magnetic Drug Targeting (MDT) 15

    1.3 Magnetic drug targeting in the lungs 18

    1.4 Chapters overview 20

2 Nanomagnetism and further relevant properties of magnetic nanoparticles 21

    2.1 Magnetism at the nanoscale and related MNP properties 21
        2.1.1 Superparamagnetism: 22
        2.1.2 The zeta potential 27
        2.1.3 The Cluster Theory 27
        2.1.4 The nanoparticle shape 30

    2.2 Toxicity Vs. biocompatibility 32

    2.3 More efficient transport systems in MDT 37

3 Magnetic guiding of nanoparticles for drug targeting in the blood vessels 39

    3.1 Modeling and Magnet 39
        3.1.1 Modeling the MDT process 39
        3.1.2 The magnet 45

    3.2 Animal experimentation and results 52
        3.2.1 Targeting of microbubbles to the cutaneous blood vessels of a mouse (the in vivo application) 52
        Animal selection: 53
        Animal preparation: 53
        3.2.2 Results: 56
        Magnetic trapping of Microbubbles in the microvasculature of the dorsal skin: ...57

4 Magnetically guided nanoparticles for drug targeting in the lungs 61

    4.1 Physiology of the human lungs 62

    4.2 Simulating the Lung Drug Targeting procedure 65
        4.2.1 Implemented model geometry for the simulation 67
        4.2.2 Magnetic forces and particle trajectories 70

- 4.2.3 The effects of intubation 80
- 4.2.4 Pre-clinical evaluation of Lung Drug Targeting 81
- 4.3 Breath-synchronous lung drug targeting 85

5 Cellular uptake of nanoparticles following successful targeting 87

- 5.1 Enhancing cell permeability with static magnetic fields: Magnetofection™ 88
- 5.2 Cell transfection with dynamic magnetic fields 91
  - 5.2.1 Rotational magnetic system and pulsating fields for cell transfection 92
  - 5.2.2 Experiments 97
  - 5.2.3 Results 101

6 Discussion and conclusions 111

7 Outlook 114

List of Tables 116

List of Figures 117

List of References 122

Abstract

„Remember, all we are trying to do is kill the cancer faster than we kill the patient! "
James F. Leary, Ph.D.
Endowed Professor of Nanomedicine
Full Professor of Basic Medical Sciences, School of Veterinary Medicine
Purdue University

Although medicine has made great progress in the last centuries and decades, it is still facing basic challenges that make doctors fail to efficiently and successfully treat the continuously emerging diseases and ailments due to ageing, industrialization, pollution and resulting biological mutations. In this context, the systemic chemotherapeutic treatment of cancer seems to be one of the most fitting examples for the wide gap between the usually followed medical approach and the theoretically optimal solution.

Extrapolating from in vitro experiments and mouse models to humans, treating children as "miniaturized" adults when analyzing therapeutic effects, estimating drug doses based on relatively coarse processes like up scaling on weight, volume or area, and flooding the human body with drugs to solely achieve a minimal effect at the ailment site are just few examples for improvement needs in medical methods.

One of the most promising approaches intended to bring more specificity and precision into the therapeutic toolbox is the directed delivery of drugs, already prophesized and described one hundred years ago by the German immunologist and Nobel Laureate in Medicine (1908) Paul Ehrlich (1854-1915) as the "magic bullet" principle. It is a visionary medical method in which active agents -such as drugs or antibodies- are guided within the human body and brought to bind directly and exclusively to their biological target. This approach was triggered and has been remarkably promoted by the introduction and continuous development of nano-sized medical systems since the 1950s, and is expected to experience a real breakthrough by the clinical validation of the so called "Magnetic Drug Targeting".

According to this technique, magnetically active nanoparticles are coated with a therapeutically active biomaterial and guided through external magnetic fields in the natural transport pathways of the body, then retained and concentrated at target sites where the biologically active load is set free. The delivered dose is augmented, side effects are lowered and the overall therapeutic efficiency is enhanced.

Especially for cancer treatment, the magnetically guided drug delivery represents a huge potential. In fact, conventional chemotherapy methods are used systemically and succeed in best cases in delivering only a fractional amount of the drug to the target sites, while the rest is absorbed by the healthy tissue of the treated body. This is so inefficient that dose levels of about 50 to 100-fold those of conventional doses need to be administered to achieve cures of cancer cells (T. A. Connors 1995). As a result, blood filtering and trafficking organs, such as the liver, the kidneys, the spleen and most importantly the heart, are the direct victims of the highly toxic substances used in chemotherapy. Even the apparently more gentle approach of applying the maximum tolerated dose at defined intervals -in order to avoid toxicity- can unintentionally lead to a chemoresistance of the tumor (C. Damyanov 2009). These shortcomings of the chemical therapy further aggravate the fact that cancer is still the worldwide deadliest disease, with an upward trend. For instance, around 25 % of all registered death cases in the European Union are reported by the World Health Organization to be caused by tumors. Despite the development of advanced anti-cancer medicine, it still remains a difficult challenge to keep costs at an affordable level. For that reason, new and more efficient cancer treatment methods with higher success rates and lower side effects and costs are urgently needed and would help physicians cope with an ever ageing world population.

In this work, we report improvements achieved in the understanding and control of the magnetically targeted drug delivery, mainly realized by the consideration of time issues and the investigation of dynamic magnetic fields. New approaches to assess the magnetic behavior of nanoparticles in suspensions as well as an advanced examination of the lung drug targeting and the mechanisms of cellular drug uptake after successful localized delivery represent the major achievements compiled in this manuscript.

The registered improvements are an important contribution to the further development of the idea of directed therapies promoted by the emerging nanomedicine. This modern medicine is expected to provide techniques that can act on a cellular and even sub-cellular level, treating ailments with considerably more accuracy.

Gradually, modern diagnostic and therapeutic techniques should elevate us slowly to the point where we can start thinking more in terms of real "regenerative" medicine. That means, we should be able to precisely and directly address pathologic tissues, save cells and organs, repair and heal them, rather than extinguish them.

Zusammenfassung

Zur Trauer von Paul Ehrlich, dem bedeutendsten deutschen Immunologen, schrieb Kaiser Wilhelm II. in seinem Beileidstelegramm: „Ich beklage mit der gesamten gebildeten Welt den Tod dieses um die medizinische Wissenschaft und die leidende Menschheit so hochverdienten Forschers, dessen Lebenswerk ihm bei der Mit- und Nachwelt unvergänglichen Ruhm und Dank sichert." (Wikipedia 27. Mai 2010)
Mehr als hundert Jahre nach Ehrlichs Tod verfolgt die "Nachwelt" noch mit großen Schritten eine seiner wichtigsten Visionen, die er während seiner Arbeiten zur Behandlung der Syphilis entwickelte: eine „Zauberkugel" (magic bullet), die einen gegebenen krankmachenden Erreger gezielt abtöten kann.

Ganz nach diesem noch -mehr denn je- aktuellen Prinzip, entwickeln Forscher heutzutage weltweit neue Methoden, um nicht nur Krankheitserreger, sondern auch befallene Gewebe, spezifisch zu behandeln. In den letzten Jahren entwickelte sich dadurch die Medizin von der konventionellen Anwendung, über die personalisierte Behandlung, wo die genetische Information eines jeden Patienten präventiv untersucht werden kann und die Ergebnisse zur Auswahl und Anpassung der Therapie-Art herangezogen werden, bis hin zur "Nanomedizin", einer neuen Ära der Arzneimittel-Konzipierung, -Synthese, -Dosierung und -Verabreichung, die Therapien auf zellulärer und sub-zellulärer Ebene ermöglichen sollte.
Mediziner sind heutzutage weit entfernt von der Darstellung von Christian Friedrich Hebbel (18.03.1813 - 13.12.1863), dass "ein Arzt eine Aufgabe hat, als ob ein Mensch in einem dunklen Zimmer in einem Buche lesen sollte". Sie sind in der Lage, durch die Integration der Nanotechnologie im biomedizinischen Bereich, Gewebe und Zellen, die durchschnittliche Dimensionen von 10 µm haben, mit Nanosystemen im Submikrometer-Bereich zu adressieren und gezielt zu behandeln.

In diesem Rahmen präsentiert sich das Magnetic Drug Targeting (MDT) als besonders wirksamer Therapie-Ansatz. Dabei werden Wirkstoff-beladene magnetische Nanopartikel über externe Magnetfelder im Körper geführt und an einem gegebenen Krankheitsort lokal angereichert. Die verabreichte Wirkdosis wird dadurch erhöht, Nebeneffekte minimiert. Besonders in der Krebsbekämpfung verspricht dieser Ansatz hohe Erfolgsquoten und eine Reduzierung der ohnehin enormen Chemo- und Radiotherapie-Kosten, die meistens einen bremsenden Effekt auf die Entwicklung und Verbreitung zahlreicher Behandlungsmethoden haben. An dieser Stelle sei daran erinnert, dass Krebs nach wie vor die weltweit wichtigste

Todesursache ist, an der schätzungsweise 11.5 Millionen Weltbewohner im Jahre 2030 sterben werden, was einem Anstieg von 45% zum Jahre 2007 darstellt.

Die zielgerichtete Arzneimittel-Applikation, zu Englisch "Directed Drug Delivery", soll hierfür Lösungen anbieten, die Tumore spezifisch angreifen und ausschalten können. Durch eine magnetische Lenkung und Anreicherung wird dieses Verfahren weiter optimiert. Die somit entstehende MDT-Methode eignet sich für Anwendungen in der Blutbahn, sowie in den Atemwegen von Patienten, mit entsprechenden Anpassungen. Entscheidend ist hierbei vor Allem das eingesetzte Magnetfeld, in Bezug auf Amplitude, Homogenität und Dynamik. In zahlreichen wissenschaftlichen Arbeiten, wurden bisher Erfolg versprechende Ergebnisse präsentiert, die überwiegend durch die Manipulation und Aufkonzentrierung von Nanopartikel-Wirkstoff-Komplexen mit statischen Magnetfeldern realisiert wurden. Eine hierzu komplementäre Betrachtung mit dynamischen Magnetfeldern wird in dieser Arbeit untersucht.

Im Rahmen dieses Forschungsprojekts wurden Ansätze mit statischen und dynamischen Magnetfeldern zur Verbesserung des Magnetic Drug Targeting theoretisch überprüft, simulativ validiert und systemtechnisch umgesetzt. Nach einer ausführlichen Untersuchung der Nanopartikel-Eigenschaften, die den MDT-Effekt überhaupt ermöglichen und besonders beeinflussen, wurde der Anreicherungsprozess unter Magnetkraftwirkung modelliert und ein für Anwendungen in der Blutbahn optimiertes Magnetsystem simuliert, konstruiert und bei in-vivo-Versuchen eingesetzt. Dadurch konnte eine aktive und vor Allem reproduzierbare Retention von beladenen Nanopartikel-Komplexen in den Arterien und Venen der Rückenhaut einer Maus verzeichnet werden. Analog zur Anwendung in den Blutgefäßen wurde anschließend das MDT in den Atemwegen untersucht. Hierbei wurden Magnetfeldrechnungen mit strömungsmechanischen Simulationen kombiniert, um das Verhalten von magnetischen Aerosolen im oberen Bereich der Lunge zu beschreiben und die entsprechenden Parameterwerte für deren gewünschte Sedimentation zu identifizieren.

Es folgten in-vitro-Experimente in Kooperation mit dem Klinikum rechts der Isar, in denen gezeigt wurde, dass dynamische Magnetfelder die Effizienz der Einschleusung von Genmaterial ins Zellinnere (Transfektion) signifikant erhöhen.

Die erzielten Ergebnisse, von der erreichten, theoretischen Verständnistiefe, über die gerätetechnischen Implementierungen, bis hin zur experimentell realisierten Steigerung der Therapieeffizienz in Targeting und Transfektion bestätigen das Potential des "Magnetic Drug Delivery"-Ansatzes und stärken die in den letzten 30 Jahren langsam gewachsenen Hoffnungen auf wirksamere und gezieltere Therapie-Methoden, insbesondere gegen Krebs.

1 Background and Motivation

Along with the progressive evolution from conventional to personalized medicine, hopes have emerged, that a further step towards more specific and efficient therapies can be achieved. Beyond treating patients in a differentiated way based on their individual responsiveness to drugs, which can be assessed through fine molecular tests or tests on relatives of the patient to establish hereditary (genetically determined) predispositions to certain diseases, it is believed that more selective methods even within an individual can be implemented. The most advanced vision of this vision is the concept of single-cell medicine, also termed "Nanomedicine", which is gaining in importance and promising advanced medical interventions on cellular and sub-cellular or molecular level.

Despite the continuous adaption and integration of great technologies into healthcare processes, the established traditional methods physicians currently follow to treat various maladies are still suboptimal, as they are very often based on rather coarse estimations. For instance, up scaling drug doses based on the weight or volume of patients is a standard medical method, completely lacking in precision or specificity (Sharyn D. Baker, 2002). In addition to that, nowadays' therapies are rather meant to eradicate diseased tissues or cells and eliminate them. In doing so, the affected cells are forced to undergo necrosis, which is an induced death process where cells lose their membrane consistency and spill their content into the cellular milieu (Helmtrud I. Roach 1999). Considering the fact that cells naturally possess various ways to investigate and sample their environment by internalizing small quantities of the surrounding material found in the extracellular space and dissolving them, or the fact that cells are able to enclose and degrade dangerous intrusions, it becomes evident that cells dispose of powerful chemical substances (enzymes) and tools (lysosomes) to attack biological entities and disintegrate them (W. E. Walker 1968). These harmful agents are set free to the neighboring tissue, if the containing cells are damaged or caused to die in a different way than safe apoptosis (Tarl W. Prow 2004). It then quickly comes to a lack of chemical signals to the immune system, which triggers a cascade of destructive events in the neighboring tissue. This leads to more damage, especially on the side of the still safe bystander cells. It is thus of great interest to pursue new methods through which a safer "switching off" or elimination of the diseased tissue is induced. This happens, when specific treatment, preferably in the same size range as cells and organelles, is achieved. For this purpose, submicron systems that are controllably synthesized down to the 5-10 nm range, and therefore approach even the size of single proteins, are needed for a very precise and targeted treatment.

On account of this, next generation nanomedicine technologies are focusing on such nanosized tools and solutions capable of even detecting diseased cells, most importantly in tumoral tissues, infiltrating their cytomembrane and performing directed intracellular repair operations or triggering an apoptosis process (Marco A. Zarbin, 2010).

Besides improved drug delivery, the introduction of nanobiotechnologies as well as nanoscale-structured machines and devices into medical practice generally holds great promise with regard to better diagnostics, transplantation procedures, better implants, real regenerative medicine, and minimally invasive surgery using nanorobots or in combination with adapted catheters (Jain, 2008).

Especially in the field of treating cancer which is one of the leading causes of death in the developed countries, the nanotechnology based approach represents a promising tool to add more specificity to the ordinary treatment.

As a matter of fact, standard chemotherapy which is one of the oncological treatments applied if the addressed tumors are not resectable can remarkably be improved and directed through nanosystems. Because it aims at delivering cytotoxic substances to the rapidly dividing cancerous cells in order to kill them, this systemically applied method also affects and harms other cells of the organism that are under normal circumstances fast-dividing. Due to this undesired side effect, not only cells of the bone marrow, of the lining of the digestive tract and in the hair roots are attacked, but also entire organs, mainly responsible for the circulation and filtration of blood, such as the heart, the liver and the kidneys, are damaged.

By flooding the patient with chemotherapeutic drugs, clinicians achieve rather non-significant to small effects, deploy a series of unwanted side effects that unnecessarily involve and stress the whole body, and cause exorbitant expenses in the healthcare system (Plank, 2009).

In this scope, it is believed, that nanomedical systems would not only enhance treatment efficiency but also help minimize or even eliminate side effects and reduce costs. Srinivas et al mentioned that using nanotechnology in medicine can be an important complement to existing technologies and make a tremendous contribution to cancer detection, prevention, diagnosis, and treatment (Pothur R Srinivas, 2002). Especially the early detection of tumors would help treat them more efficiently, before they undergo mutations and gain resistance to drugs. With regard to that, nanomaterials, such as quantum dots, gold nanoparticles, magnetic nanoparticles, carbon nanotubes, and gold nanowires, exhibit unique physical, optical and electrical properties that in combination with proteins, antibody fragments, DNA and RNA fragments which are the base of cancer biomarkers have proven to be very useful in cancer sensing and monitoring (Young-Eun Choi, 2010). For the treatment of cancer, nanosystems can be used as drug delivery devices where the drugs are embedded in a nanocarrier with functionalities specific only to the cell in question, or they can be utilized to heat up tumoral tissue or to perform gene therapy. For instance, over 100 drugs based on nanotechnology have been in development for the last years, with some having been already approved, e.g. Doxil (a liposome preparation of doxorubicin) and Abraxane (paclitaxel in nanoparticle formulation) (Jain, 2008).

As shown in several studies (Jain, 2008), the most suitable nanosystem candidates to facilitate drug delivery and, more generally, achieve the goals of nanomedicine are multilayered nanoparticles. These are submicron beads having at least one dimension ranging from 1 to 100 nm and exhibiting size-dependent physical (optical, electrical or magnetic) and chemical properties remarkably different from those of bigger samples made of the same materials. It is here noteworthy, that the radical change in properties is just a natural result of the matter being shrunk to nanoscale.

In fact, just by attentively observing nature and its "natural nanoparticles", like viruses, it becomes quickly obvious that modern science has to mimic these structures to have access to a toolbox on cellular and organelle scale, without causing severe interference with the organism (Benyus 1997) (Salata 2004). It is even believed that nanoparticles can interact with the components of cells such as cellular membranes and nucleic acids which are nanometer in scale (Pascal R. Leroueil, 2007).

Nanoparticles have been intensively studied and developed over the last few decades, especially as biodegradable carriers for drug delivery applications. The carried therapeutics can be entrapped, adsorbed, attached, dissolved or encapsulated into the nanoparticle, leading also to new constructs, such as nanospheres and nanocapsules. Depending on the form used, the release process of the transported biological agents varies. The advantage of these nano-scale objects is, first, that they are small enough to penetrate deeper porous structures in the organism such as smaller capillaries and even cell membranes. When taken up by cells, they can deposit their load in the intracellular milieu and assure an efficient drug accumulation at target sites. Their small size, if precisely controlled, gives them furthermore sharp optical properties so that they can act as very efficient fluorescent probes, for example by emitting narrow light (Salata, 2004). Second, the coating of nanoparticles with appropriate biodegradable materials allows not only for biocompatibility but also for a controlled and slow release of the carried drug, as many cancer treatments require that only the targeted organ receives the drugs at a pre-programmed rate and at well-defined concentrations (Sanjeeb K. Sahoo, 2003). Therefore, it is very important to have nanoparticles with tightly controlled, narrowly distributed sizes and accordingly engineered coatings.
But size and hull are not the only problems. In fact, having tackled and solved the issue of miniaturization, it still remains a challenge to handle, steer and control the

nanosystems within the body and to have them perform the intended tasks with precise mechanisms of action at the relevant sites. This is achieved by passive or active targeting. In passive targeting, the nanoparticle coupled to the therapeutic agent passively reaches the target organ by making use of the enhanced permeation and retention effect (EPR) involving the fast growth of tumoral blood vessels that are in consequence porous and abnormally fenestrated (Wayne L. Monsky, 1999). This way of passing pores and cell membranes is however nonselective and based only on size criteria (Pascal R. Leroueil, 2007). Active targeting, in the contrary, is achieved by conjugating the carrying nanoparticle or the carried drug to a cell or tissue-specific ligand that only binds at the target site (ALF LAMPRECHT, 2001). As a complement to these methods, there has been gradual interest in the last years to investigate and introduce a further targeting technique, involving magnetic nanoparticles that can be magnetically directed to target sites.

1.1 Magnetic nanoparticles in medicine

Besides nanorods, nanotubes (Sang Jun Son, 2005), nanowires, cylindrical and plate-like shaped nanoassemblies, which are all nanosized structures and constructs used in biology and medicine, spherical nanoparticles represent the largest category of biomedical nanosystems (Salata, 2004) (Jain 2008). These are solid or colloidal particles consisting of macromolecular substances and having a diameter that ranges from a few to 250 nm (Torchilin 2006). Their spherical shape is often intrinsically defined through the synthesis process and is most suitable for loading additional layers to the particulate, such as functional groups, and for meticulous operations such as pore passing and cell penetration. Nanoparticles are used in a variety of biomedical applications, especially as fluorescent biological labels thanks to their protein-like size making them suitable for bio-tagging and labeling, bio-detection of pathogens and proteins, tissue engineering, separation and purification of biological molecules, drug and gene delivery (Hongwei Gu, 2006), cancer therapy, or as contrast enhancers for medical imaging, such as gold nanoparticles in x-ray imaging (AuroVist, 2009) or superparamagnetic iron oxide nanoparticles (SPIONs) in magnetic resonance imaging (Claire Wilhelm, 2008) (Hyon Bin Na, 2007). Evidence has also been furnished that nanoparticles, if drug-loaded and conjugated to the surfaces of therapeutic cells, could enhance cell therapy (Matthias T Stephan, 2010). But in most cases, the nanoparticle is used as a carrier of a bioactive load and therefore acts as an effective drug delivery device. In order to selectively detect the target and securely interact with its biological entities, a nanoparticle should bear a molecular or biological coating consisting of functional layers, sensing extremities and a biocompatibility envelope. Figure 1 depicts the composition of a typical nanoparticle system used in biomedical applications and exhibiting diverse options for chemical decoration with various functional molecules.

In the context of nanomedicine and its approaches for directed therapy, fitting nanoparticles with additional magnetic properties presents a further feature that dramatically enhances their ability to specifically target entities in the body. Systems comprising such an "intelligent" particle core enable the possibility of addressing, distally guiding and concentrating drug carriers through external magnetic fields. Combined with a receptor recognition layer and a therapeutic load, such a nanoengineered device can be considered as the tool of the future with the highest potential for therapeutic applications.

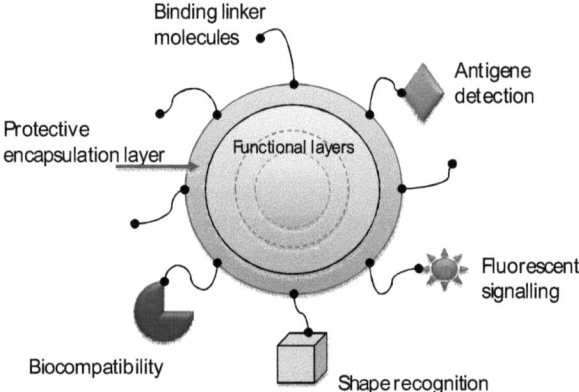

Figure 1: Composition of a standard nanoparticle system for biomedical applications (adapted from (Salata 2004))

1.1.1 Manufacturing:

Nanosized materials can be obtained by two approaches. When opting for a top-down technique, bulk material is gradually made smaller through physical processes including crushing, grinding and milling. Larger particles are thus broken up into smaller entities, until the nanoscale is reached.

In the bottom-up approach, the nanosized material is built starting from the atomic level by wet phase chemistry and defined sizes are reached by stopping the corresponding chemical reaction at exact times to freeze the particle growth process (Halim, 2008).

Magnetic nanoparticles usually consist of components exhibiting magnetic properties, like iron oxides (such as Fe_3O_4 and $\gamma\text{-}Fe_2O_3$), pure metals (such as Ferrum Cobalt and nickel), their chemical compounds and alloys.

The most important and established methods of magnetic nanoparticle synthesis are:

- Co-precipitation

A simple and efficient technique to build iron oxides through adding a base to an aqueous Fe^{2+}/Fe^{3+} salt solution at room temperature (or above) and under conditions of inert atmosphere (An-Hui Lu, 2007).

- Thermal decomposition

Heat is used to cause chemical decomposition of organometallic compounds into monodisperse magnetic nanocrystals. This is achieved in the presence of organic solvents and stabilizing surfactants. Shape and morphology of resulting crystals are mainly controlled through the ratios of the used reagents, reaction temperature and time (An-Hui Lu, 2007).

- Microemulsion

As microemulsions are stable dispersions of two liquids, they can be used to synthesize nanoparticles through precipitation caused by the addition of solvents such as ethanol or acetone. The resulting nanoparticle-precipitate can be isolated through centrifugation or filtering. Examples of particulates obtained by microemulsion are gold-coated cobalt/platinum, metallic cobalt and cobalt/platinum alloys (An-Hui Lu, 2007).

- Hydrothermal synthesis

Magnetic nanoparticles, in particular iron oxides, can also be synthesized by oxidation of iron containing chloride compounds in basic aqueous solutions under elevated pressure and temperature (Song Ge, 2009) (Yuhong Mi, 2006).

The different synthesis methods for magnetic nanoparticles are depicted in Table 1, with an emphasis on the major factors of feasibility and scalability.

Synthetic method	Synthesis	Reaction temp. [°C]	React. time	Solvent	Surface-capping agents	Size distribution	Shape control	Yield
co-precipitation	very simple, ambient conditions	20-90	minutes	water	needed, added during or after reaction	relatively narrow	not good	high/ scalable
thermal decomposition	complicated, inert atmosphere	100-230	hours-days	organic compound	needed, added during reaction	very narrow	very good	high/ scalable
microemulsion	complicated, ambient conditions	20-50	hours	organic compound	needed, added during reaction	relatively narrow	Good	low
hydrothermal synthesis	simple, high pressure	220	hours, ca. days	water-ethanol	needed, added during reaction	very narrow	very good	medium

Table 1: Comparison of the major established methods of magnetic nanoparticle synthesis (Song Ge, 2009).

For biomedical applications, magnetic nanoparticles are generally fabricated along "bottom-up" processes, starting from atoms or molecules and growing the nanostructures in a controlled way (Arruebo Manuel 2007). Such particles commonly consist of a metal or metallic oxide core serving as magnetic element such as iron, nickel and cobalt and their chemical compounds. As a synthesis technique, co-precipitation is most used, as it is a very efficient and easy to scale-up process.

Theoretically, all magnetically addressable components can be used to fabricate magnetic nanoparticles for clinical use. Moreover, the general rule applies that the stronger the responsiveness to magnetic attraction, the better the control over the particulate. This led scientists to the use of cobalt and pure iron, or alloys of iron, platinum, cobalt and carbon to synthesize particles showing strong reaction to magnetic fields and higher saturation magnetizations (Jana Chomoucka, 2010). However, a major problem related to these materials refrained clinicians from applying them in biomedical purposes. For instance, cobalt is highly toxic and pure iron is hyper-sensitive to oxidation, which imposes the encapsulation of particulates made of these materials in inorganic coatings to prevent or minimize toxicity or oxidation, respectively. Due to these constraints and with regard to the simplicity of their synthesis, iron oxide magnetic nanoparticles have advanced to the status of the most commonly used magnetic nanoparticles. Superparamagnetic magnetite (Fe_3O_4) and maghemite (γ-Fe_2O_3) can easily be fabricated through physical or chemical processes, and are under controlled conditions totally bio-compatible (Torchilin, 2006).

Precipitation of Fe(II)/Fe(III) hydroxide in an inert atmosphere, followed by adequate stabilization (Mykhaylyk Olga, 2007), produces magnetite particles that are easily dispersible in water, having generally ellipsoidal shapes and sizes between 3 and 30 nm with a size distribution of 10 to 20% (Torchilin, 2006). Similar to the established industrial techniques available to use and manipulate magnetic nanoparticles in suspensions, such as in audio speakers/boxes or magnetic separation for the purification of industrial water (Franzreb, 2003), the obtained iron oxide nanoparticles can be sorted to narrow their size distribution to about 5% (Torchilin, 2006).

When used clinically, magnetic nanoparticles need to meet all necessary biocompatibility criteria and have to fulfill their biomedical purpose. For that, special coatings have been developed assuring non-toxicity, stability and molecular reactivity, and thus enhancing physical and chemical properties of the nano-systems. Additionally, magnetic interaction between particles imposes the intercalation of layers encapsulating the single beads and avoiding unwanted agglomeration. To incorporate this concept, modern techniques to synthesize magnetic nanoparticles are mostly based on a core/shell structure. The core being the iron oxide component, the shell represents its coating and enables better dispersion and stability against oxidation, as well as the binding of drug loads. Special attention should however be brought to size, as the total diameter of the particulate should be below 100 nm to avoid fast clearance by the reticuloendothelial system (RES).

The surface coating of magnetic nanoparticles is achieved through layers of polymers, proteins, silica or other organic and inorganic materials (dextran, starch, polyethylene glycol (PEG), polyvinyl alcohol (PVA), gold, etc.) (Jana Chomoucka, 2010) (Torchilin, 2006). In a layer-by-layer building approach, the nano-system is then enrobed in further functionalization (i.e. anchoring of biofunctional molecules onto the nanoparticle) shells protecting it from degradation and enabling its activation and bio-interaction capabilities (An-Hui Lu, 2007).

1.1.2 Biomedical applications of Magnetic Nanoparticles:

Magnetizable nanoparticles present highly interesting properties allowing for multiple medical and biological applications. For instance, researchers use MNPs for:
- cellular and macromolecular separation processes (bioseparation), where target cells or molecules labelled with nanoparticles are isolated from a mixture with high specificity and sensitivity (José Luis Corchero, 2007) (Hongwei Gu K. X., 2006)
- contrast enhancement in diagnostic and interventional imaging, especially magnetic resonance tomography, where the MNPs used are saturated in the scanner field and create a perturbing dipole that shortens the T_2 relaxation time of protons more that T_1, thus contributing to the contrast of the acquired image (Pankhurst Q A, 2003).
- magnetically mediated hyperthermia, where magnetic nanoparticles that are exposed to an alternating magnetic field are heated to reach temperatures above 43°C, thus selectively destroying cancer cells that are more sensitive to heat than healthy tissue. This selective inducting heating of targeted regions in the body might imply other fabrication techniques for the nanoparticles (José Luis Corchero, 2007) (Torchilin, 2006).
- arterial embolization therapies where nanoparticles are used to block blood vessels and delay tumor growth or inhibit tumor angiogenesis
- gene transfer and therapy, where magnetic nanoparticles are used to introduce foreign genetic material into target cells, thus increasing the expression of a given gene or allowing for genetic engineering operations (José Luis Corchero, 2007)
- for targeted drug delivery and release, especially in cancer treatment, where drugs associated to magnetic nanoparticles are brought to accumulate at a desired site in the organism under magnetic focusing, thus enabling higher concentrations of bio-active substances at targeted tissue locations (Torchilin, 2006)

In these application fields as well as the newer developments of magnetic nanoparticle usage, the administration of the nano-systems occurs systemically via the blood through intravenous or intra-arterial administration, subcutaneously and or directly through injection into diseased organs or tumor tissue (Arruebo Manuel, 2007), as well as over inhalation (oral administration) to the inner parts of the lungs.

1.2 Magnetic Drug Targeting (MDT)

"Was mich nicht umbringt, macht mich stärker"

("That which does not kill me makes me stronger")

Friedrich Nietzsche (1844 – 1900), Sprüche und Pfeile, 8.

Despite the enormous advances made in the development of pharmaceuticals and novel drug delivery methods, the demand for more efficient and financially bearable treatments for a multitude of disorders, like cardiovascular diseases, ailments of the respiratory and nervous system as well as tumors, is still present and expected to grow (Sanjeeb K. Sahoo, 2003). Normal ways of administrating drugs imply a uniform distribution of that drug in the body of the treated patient as it is injected in the blood and systemically delivered to all vascularized organs. This inevitably leads to the pharmaceutical substance distributed at other sites than the therapeutic site, which decreases the efficiency of therapy, raises treatment costs -as a larger drug quantity is required to deliver the needed effect at target site- and may cause toxic side effects at healthy tissue level (Yokoyama, 2005). Especially in chemotherapy, these limitations are observed and represent a major challenge to reach full treatment success.

In fact, tumors can quickly develop drug resistance and if treated in a sub-optimal way, remaining tumoral tissue can rebuild to a more resistant malignancy, imposing therefore a change in the therapy strategy. Moreover, the organism irreversibly changes after it is treated with a given medicine, so that it will not react to further drugs the way it would have done before alteration. Each and every administration of a new therapeutic substance has an impact on the responsiveness of the organism to the next steps of a medication plan. Therefore, any drug therapy should be planned and carefully conducted till the complete elimination of the ailment.

With regard to these challenges, Magnetic Drug Targeting (MDT) has emerged as a concept of concentrating drugs at therapeutic sites through binding them to magnetic nanoparticles that are guided in the body and retained via external magnetic fields. Especially for tumor treatment, holding the chemotherapeutic agent at the desired site of activity significantly increases therapy efficacy and reduces side effects mainly observed as systemic toxicity (Torchilin, 2006).

The principle of Magnetic Drug Targeting has been examined and tested in-vitro and in-vivo, in animal models and in humans (Challa S.SR. Kumar, 2005), since the 1970s. The most relevant results have been so far the successes achieved by Widder et al. in 1979 when they reported a 200-fold increase -compared to systemic intravenous administration- in the concentration of the chemotherapeutic drug doxorubicin retained at a targeted tumor area in a rat's tail. They had opted for an intra-arterial application (Sophie Laurent, 2011). In 1981, the same group also achieved a complete tumor remission in 77% of a colony of Yoshida rats having sarcomas in their tales (Torchilin, 2006). Many research groups followed and reported successful applications in hamsters, rabbits and swine. In 1996, Lübbe et al. successfully treated tumors in mice and rats through an intravenous injection of 100 nm sized magnetic nanoparticles coated with anhydroglucose linked to 4-epirubicin (an analog of doxorubicin with similar anti-tumor activity but significantly lower cardiotoxicity). They used a magnetic field of 0.2-0.5 T and achieved significant drug accumulation at target-site. The injected nanoparticles landed finally mostly in liver and spleen, and only minor amounts deposited in heart, lung and kidney. Also in 1996, Lübbe et al. reported a first phase I clinical trial involving 14 patients with squamous carcinoma (a cancer of a kind of epithelial cell) of the breast or head and neck. They received intravenous injections of 4-epirubicin coating 100 nm ferrofluidic particles and were exposed, at the relevant target regions, to a 0.5-0.8 T magnetic field for 60-120 minutes. The drug accumulated in the tumor area in six patients with no rejection. Systemic effects were reduced and toxicity was lower than after 4-epirubicin treatment alone (Challa S.S.R. Kumar, 2005). MRI examination revealed, however, that more than half of the magnetic nanoparticles landed in the liver (Torchilin, 2006).

In 2000, Alexiou et al. successfully treated New Zealand White rabbits with implanted experimental squamous carcinoma in the hind limb. Magnetic nanoparticles bearing mitoxantrone were intravenously (ear vein) and intra-arterially (femoral artery) injected and retained at tumor site through external magnetic fields, resulting thus in complete and permanent remission of the tumor (Christoph Alexiou, 2000). Further successful studies in rabbits (R Jurgons, 2006) and hamsters (Kubo T, 2001) have also been reported in 2006 and 2001 respectively.

It is at this stage also noteworthy to mention the positive results reported by the company FeRx in 2000 and 2002 and indicating a successful use of carbon-coated iron particles bearing doxorubicin and having diameters of 0.5-5 μm in a clinical trial targeting inoperable liver cancer cells. Here too, therapy-related toxicity has been reported as very low (Torchilin, 2006).

Despite the differences in nanoparticle-drug-complexes used, the magnetic systems applied and the various ways of administration chosen in these studies, a common denominator for Magnetic Drug Targeting in the blood vessels remains the need for an accessible tumor that is enough vascularized and irrigated by blood, in order to allow nanoparticles to reach it. It is also a challenge to appropriately tune all parameters and setup requirements to achieve best results. In general, the efficacy of MDT treatment depends on the nanoparticle properties and behavior, the magnetic field, and the blood flow conditions.

In a relevant literature note, the specifications of a successful setting indicate a needed high magnetization of the MNPs in order for them to be sufficiently addressed by the external magnetic field and overcome linear blood flow rates of 10 cm/s in arteries and 0.05 cm/s in capillaries. Nanoparticles should also fulfill biocompatibility requirements, exhibit sizes below 200 nm, anti-agglomeration behavior as well as a long circulation capability (Challa S.S.R. Kumar, 2005). Kumar et al. also suggest the used magnetic field has to have a magnetic flux density of at least 0.8 T for nanoparticulates featuring 20% of magnetite. For the majority of nanobeads, this density might be as low as 0.2 T but with a field gradient of 8 T/m for arterial application. The maximum depth that can be reached is, in that case, 8-12 cm (Challa S.S.R. Kumar, 2005).

In fact, the weak magnetic fields generated by the available systems result in a restriction of the applications of magnetic drug targeting. For instance, the limited depth reached by the magnetic field makes it rather impossible to treat ailments at tissues or organs deep inside the human body. Therefore, MDT -as it is allowed by the available magnet technology- is more suitable for pathologies near the surface of the body, where a magnetic system is easily placed and the distance to the transportation pathways through which particles flow is minimized. Examples for these diseases which are treatable under similar circumstances are surface tumors like prostate and breast cancers or skin malignancies.

But depth is not the only challenge facing MDT. Putting aside the potential hazard of nanoparticle dislocation to lungs and brain, if MNPs passed the blood-brain barrier, major issues still remain at the level of the particles, their stability, their magnetic responsiveness, toxicity, drug conjugation, magnet technology, and flow control. Another crucial point is the effect of "mis-targeting" which leads mainly to the damage and loss of "innocent" bystander cells, and in the most severe cases enables non-linear reactions of the organism that are very hard to contain or to counter.

The need for concentrating and intelligently guiding the therapeutic effect is therefore obvious and becomes more important if we additionally consider avoiding the costly drug wastage related to systemic application. This is exactly the area of activity of MDT which should be further developed to allow for solutions that enable custom-designed treatments according to the needs of each and every patient.

1.3 Magnetic drug targeting in the lungs

Just as magnetic drug targeting makes use of the natural –and in case of tumors modified- vasculature and blood circulation to efficiently deliver chemotherapeutics to specific compartments of the organism, it is intuitively imaginable to expand this "nanotechnology approach to drug delivery" to other circulation systems in the body, for instance, to the respiration system. This has led in the last years to the exploration of a new area of MDT, the so called "Lung Drug Targeting (LDT)". The idea behind it is simply to apply the experiences and results gained from magnetic drug targeting in the blood vessels to assure a local treatment of pulmonary diseases (Dames, 2007).

This application of MDT gains in importance when we recall the fact that lung cancer causes the death of 1.3 million people worldwide, representing thus the most fatal cancer type for men and the second deadliest for women (Dahmani Ch., 2008). Lung drug targeting emerges here as a promising technique to treat carcinomas in the lungs, by delivering drugs through the pulmonary ways directly to the tumor sites in the trachea and bronchial tree over actively loaded superparamagnetic nanoparticles. These particles are encapsulated in aerosols and guided by external magnetic fields.

It is here noteworthy to mention that aerosol application –on which LDT is based- is an established, non-invasive route of drug delivery to deep pulmonary compartments, mainly to the alveoli, and which is routinely used in the medical treatment of many diseases such as asthma, COPD or mucoviscidose (Gonda, 1990) (G. Scheuch, 2006). This is mainly due to the fact that delivering medications by inhaling an aerosol has intrinsically important advantages over systemic drug delivery, such as the direct deposition of medication to airway receptor sites, the fast onset of action of the deposited drugs, the lower systemic bioavailability of medications administered via inhalation and the ease and convenience of self-administration by patients. Furthermore, aerosol-based drug delivery allows for lower medication dosages and thus for cost reduction, while achieving the desired therapeutic effect.
Making use of these same advantages, the intratracheal administration route is nowadays being intensively evaluated for tumour healing (L. W. Wattenberg, 2004) (Y. Zou, 2007).

LDT successes have been reported by P. Dames et al in 2007, when they showed in a simplified computer-based simulation of a theoretically elaborated concept, and in experiments performed on mice, that targeted aerosol delivery to the lung is feasible, with aerosol droplets that contain superparamagnetic iron oxide nanoparticles. For their calculations, P. Dames et al involved magnetic gradients of over 100 T/m and spherical aerosols of 3.5 μm comprising MNPs with a 50 nm diameter (Dames, 2007).

In 2005, Ally et al demonstrated –based on an in vitro model- that magnetically targeted deposition of aerosols presents a feasible, potential technique to treat cancers of the lungs (Javed Ally, 2005).

Despite the evidences brought to our knowledge through these results, it can be easily noticed, that works done to explore LDT so far, still involve too simplified models, be it experimental or simulative, and achieved findings in an animal model that cannot be automatically extrapolated to human application.

Additionally to these limitations, several other aspects of LDT -of technical and clinical relevance- present important challenges and deserve thorough investigation. For instance, due to the strong dependence on the respiration flow, the exact reproduction of a given drug distribution in the lungs remains a principal issue and makes automatic comparisons of achieved results relatively complex. This is but just one simple example of the issues related to fluid dynamics that will be encountered while studying LDT. Moreover, the magnetic parameters have to be compliant with the particularities of steering magnetic nanoparticles in the inhaled air. The range of the magnetic field is with respect to this the most relevant aspect and airborne particles have to be addressed within the lungs, over distances usually inaccessible to externally applied magnetic forces, which implies the need for stronger magnets capable of generating greater field gradients. The steering is also difficult as treatment of ailments in the lungs might involve just one defined compartment of the respiration tract, making accurate particle guidance necessary to concentrate the therapeutic effect on the relevant pathways. Further synchronization of the magnetic and respiration activities is furthermore crucial to optimize nanoparticle retention efficacy.

Provided targeting and particle deposition have been successfully achieved, further challenges emerge as to the need for surmounting or avoiding the clearance mechanisms in the lungs and assuring a clinically efficient uptake of the applied drugs.

Finally, just as it is important to pre-interventionally simulate and plan the lung drug targeting procedure, it is also crucial to monitor and assess its outcomes. These multiple issues deserve special care and systematic investigation.

1.4 Chapters overview

The purpose of this work is to investigate the most relevant among the currently explored aspects of guiding drug-loaded magnetic nanoparticles in the body to achieve loco-regional targeted therapeutic effects. Hereby, both the application in blood vessels and in the pulmonary airways will be addressed.

After a thorough introduction laying the ground for a primary understanding of these two clinical applications and the motivation behind investigating them, chapter 2 will outline the nanoparticle properties relevant for a successful magnetic drug targeting, including toxicity and transportation aspects.

In chapter 3, the application of MDT in blood vessels will be examined, with a focus on magnet technology and experimental validation on animals.

The primary concern of chapter 4 will then be to explore the transition to the application in the pulmonary airways, covering simulative evaluations as well as technical and experimental aspects. Once results from both fields are presented, the common challenge of internalizing the retained drug-bearing nanoparticles at target-site will be tackled and constitutes the core of chapter 5.

Finally, in the sixth section, further relevant aspects related to the monitoring and evaluation of therapy progress, as well as imaging challenges in MDT will be covered, before results are discussed in the final chapter.

2 Nanomagnetism and further relevant properties of magnetic nanoparticles

2.1 Magnetism at the nanoscale and related MNP properties

With the lecture he gave in 1959 at the yearly meeting of the American Physical Society at the California Institute of Technology (Caltech), entitled „There's plenty of room at the bottom", the physicist Richard Feynman ignited the exploration of small-scale materials and inspired the concept of nanotechnology (Feynman, 1992). Since then, nano-scaled systems have been the focus of a continuous attention, intensively studied and investigated.

The driving force behind this interest lies in the innumerable possibilities enabled through the properties of nanomaterials that gain even further attractiveness when it comes to magnetic nanomaterials.

For magnetic drug targeting, nanosized, magnetically active particulates are used (MNPs). As these usually have an iron oxide core, they are often referred to as superparamagnetic iron oxide nanoparticles (SPIONS).

To understand how these particles, when they reach the nanoscale, exhibit extraordinary properties that totally differentiate them from bulks of the same materials, it is crucial to investigate the impact of size reduction on the surface area-to-volume ratio.

In fact, in a bulk crystal, the majority of the atoms are located in the inner part of the volume occupied by the crystal. The number of atoms within the volume exceeds by far the number of atoms on its surface. These inner atoms define –thus– the properties (physical, chemical, electrical and optical) of the bulk material.

As the size of the crystal is reduced, the surface area-to-volume ratio increases, and the influence of the atoms comprised in the surface of the material surpasses the contribution of the inner atoms to the definition of the crystal properties. This is due to the resulting larger contribution of the surface energy to the overall energy of the whole system, decreasing so the impact of the inner bulk atoms on the properties of a material (Halim, 2008).

The increasing importance of surface atoms in nanosized matter is shown in Figure 2 depicting the percentage of surface atoms in a particle against its size.

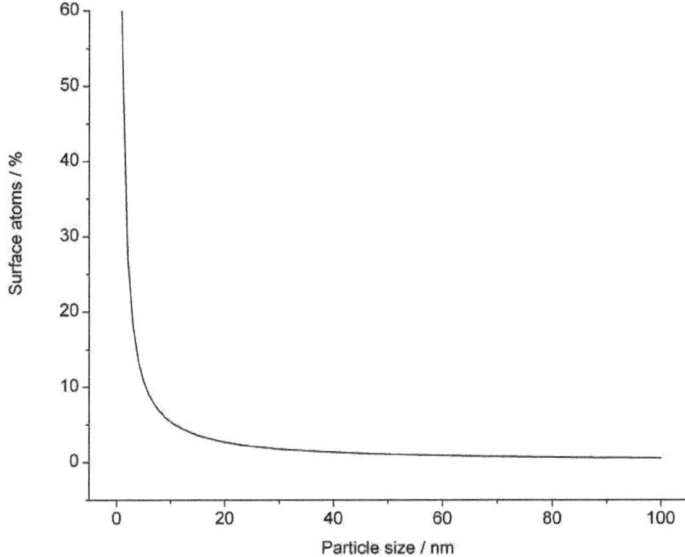

Figure 2: Relationship between the percentage of surface atoms in a particle and its size (Halim, 2008)

This clear "surface effect" is not the only aspect dominating the magnetic properties of MNPs. It is also endorsed by "finite-size effects" that give rise to the special feature of superparamagnetism which is the major property of nanoparticles making them suitable for magnetic drug targeting.

2.1.1 Superparamagnetism:

Magnetic materials owe their properties to their microstructure, comprising primarily magnetic domains and the transition regions between these domains –the so-called magnetic domain walls-. A magnetic material exhibits under certain conditions these domains which are regions within the material showing uniform magnetization: this means that in that compartment the individual magnetic moments of the atoms are aligned with one another and pointing in the same direction.

Due to energetic constraints, a magnetic material spontaneously divides into separate domains. In fact, through avoiding a state with a magnetization in the same direction all through the material, internal energy is minimized.

A material with constant magnetization over one large domain will generate a significant magnetic field reaching outside of the material volume and storing a large quantity of the magnetostatic energy. When the considered sample splits into two domains, with the magnetization of each of them in the opposite direction of the other, the stored energy within the field is reduced, as illustrated in Figure 3.

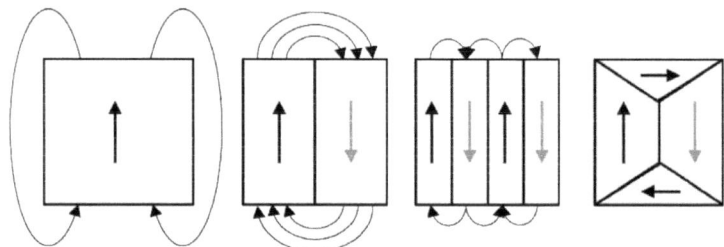

Figure 3: Minimizing magnetostatic energy stored in the external magnetic field of a material sample through the creation of internal magnetic domains. The magnetic field lines extend in loops in opposite directions through each domain, which reduces the field outside the material (Dutz, 2008).

To further minimize the field energy, each of the two domains can split into smaller parallel domains with magnetization in alternating directions, exhibiting thus smaller amounts of field outside the material.

For magnetic nanoparticles, there exists, nevertheless, a critical size below which it costs more energy to create a new domain and a new domain wall within the particle than to support the magnetostatic energy of the single-domain state. In that case, the MNP comprises a single domain (O'Handley, 2000).

This threshold size depends on the nature of the material and typically is about a few tens of nanometers (An-Hui Lu, 2007). For Fe-based nanoparticles, a single-domain structure is expected at sizes below 25 nm (Arruebo Manuel, 2007).

Table 2 depicts expected single-domain sizes of a selection of particles having spherical shapes.

Material	Critical diameter in nm
Co (hcp shaped)	15
Co (fcc shaped)	7
Fe	15
Ni	55
$SmCo_5$	750
Fe_3O_4	128

Table 2: Selected materials and the corresponding critical sizes below which they display a single-domain structure (An-Hui Lu, 2007)

In magnetic nanoparticles with single domains, the dimensions are so small that the thermal energy becomes of the same order of the magnetic energy and the spins are increasingly affected by the thermal fluctuations. As a result of these thermal fluctuations, the direction of magnetization varies randomly, similarly to the direction of the magnetic moments of atoms or ions in a paramagnet which varies under the action of thermal motion. As a consequence, the system of nanoparticles behaves, in magnetic fields and under temperature changes, like a paramagnetic material of N atoms (N being the number of NPs). The spin of each particle being free to fluctuate in response to thermal energy, the resulting coercive field is substantially equal to zero and the material exhibits no remanence, which leads to an anhysteretic B-H curve as shown in Figure 4.

The nanoparticles in this state are said to be superparamagnetic. In this state, the magnetic moment can reorient in times less than 1 ns due to thermal agitation (Sibnath Kayal, 2010).

In summary, superparamagnetic NPs are highly magnetized in the presence of an external magnetic field, but their magnetization disappears as soon as the magnetic field is removed (Sinha, 2008) (An-Hui Lu, 2007) (Arruebo Manuel, 2007).

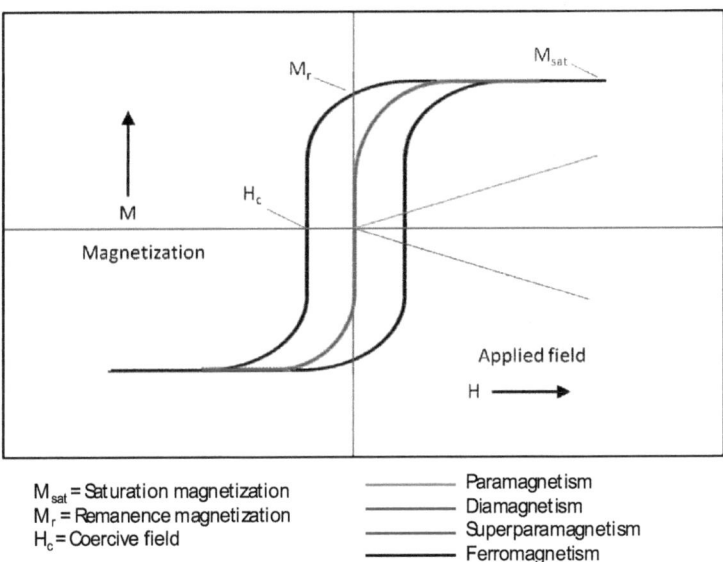

M_{sat} = Saturation magnetization
M_r = Remanence magnetization
H_c = Coercive field

——— Paramagnetism
——— Diamagnetism
——— Superparamagnetism
——— Ferromagnetism

Figure 4: Hysteresis loops for ferromagnetic and superparamagnetic nanoparticles, in comparison with para- and diamagnetic behavior.

In the work presented here, superparamagnetic nanoparticles based on iron oxide cores have been used. For the most important experiments in this study, namely the targeting in blood vessels and the transfection enhancement, so called PEI-Mag2 NPs have been selected. These are core/shell-type magnetite nanoparticles where the core has an average crystallite size of about 9 nm; their surface is coated by the fluorinated surfactant Zonyl FSA (lithium 3-[2-(perfluoroalkyl)ethylthio] propionate) combined with 25-kDa branched polyethylenimine –a type of water soluble polymer- (PEI-25Br). The hydrodynamic diameter of the particles in aqueous suspension is 28 ± 2 nm. The presence of PEI in the surface layer of the PEI-Mag2 particles results in a highly positive net ξ-potential of the particles when measured in aqueous suspension (+55.0 ± 0.7 mV). The saturation magnetisation per unit of iron weight at 298 K is 62 emu/g or $A \cdot m^2$/kg iron (Mykhaylyk O. A., 2007) (Mykhaylyk, Steingötter, Perea, Aigner, Botnar, & Plank, 2009).

Parameter	Value
Phase composition of the core	Magnetite
Mean crystallite size of the core <d> in nm	9
Saturation magnetization of the core M_s ($A \cdot m^2$/kg of iron)	62
Average iron weight per particle (g of iron per particle)	1.4×10^{-18}
Effective magnetic moment of the particle M_{eff} ($A \cdot m^2$)	8.7×10^{-20}
Iron content (g of iron / g of dry weight)	0.56
Stabilizer content (g of stabilizer / g of dry weight)	0.23
Mean hydrated particle diameter in water D_h (nm)	28 ± 2
Electrokinetic potential in water, ξ (mV)	+55.0 ± 0.7
Shell composition	$PEI-25_{Br}$ (32 mass %) ZONYL FSA (68 mass %)

Table 3: Relevant properties of the PEI-Mag2 magnetic nanoparticles

The magnetic moment of the PEI-Mag2 nanoparticles -and of every nanosystem used under magnetic locomotion- is of critical importance. It describes the magnetic response of the carriers to an imposed external magnetic field and induces the movement and migration of the particle loaded with bioactive substance in the blood, air or intracellular space.

Magnetic moments can be described using two models: magnetic poles and atomic currents. In the classical definition of magnetic moments, these are vectors pointing from the south (theoretic) pole to the north pole of a magnetic dipole and characterizing its general magnetic properties. The local value of the magnetic moment per unit volume is then defined as the magnetization of a magnetized material (J. Stöhr, 2006).

To determine the magnetic moment of the PEI-Mag2 particles, their magnetic responsiveness has been evaluated (Mykhaylyk, Steingötter, Perea, Aigner, Botnar, & Plank, 2009). H441 cells have been loaded with the PEI-Mag2 MNPs and the obtained assemblies have been exposed to a uniform magnetic field gradient of 4 T/m.

As depicted in Figure 5, two sets of 4 Ne-Fe-B permanent magnets have been placed on both sides of an optical cuvette in a spectrophotometer containing the cell suspension. The labelled cells are attracted in a direction orthogonal to the magnets. A light beam is propagated through the cuvette, perpendicular to the trajectory of the cell movement. The intensity of this light that can be sensed at a point behind the suspension becomes greater as more cells are pulled to the surface of the recipient and turbidity is diminished. The time course of the normalized turbidity of the magnetic transfection complexes is evaluated to determine the magnetophoretic mobility of the MNP-labelled cells through calculating the magnetic responsiveness in µm/s and the average magnetic moment of the cells.

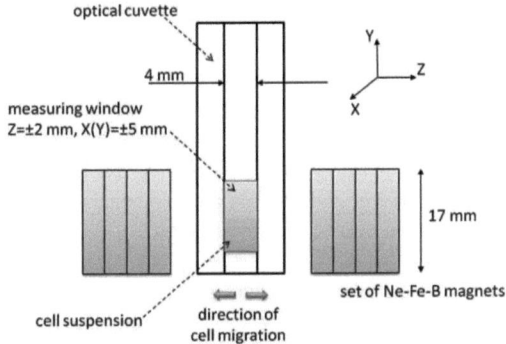

Figure 5: Measurement of the magnetic responsiveness of the magnetically labeled cells and the magnetic nanoparticles

The motion of MNPs in a magnetic field gradient obeys the formula

$$\vec{F}_m = (\vec{M} \cdot \nabla)\vec{B} \qquad (1)$$

As derived in (Leach, 2003).

$\vec{M}$ (Am²) is here the total magnetic moment of the complex containing the nanoparticles, calculated as the product of the effective magnetic moment, m_{eff}, of a single particle and the number N of MNPs attached to a given cell ($\vec{M} = N \cdot \vec{m}_{eff}$).
$\vec{B}$ (T or Vs/m²) is the magnetic flux density.
Based on the measurement of the clearance kinetics in the evaluated cell suspension, the average velocity of the labelled cell motion under the influence of

the magnetic field gradient is determined and the number of particles associated with a cell can be calculated (Mykhaylyk, Steingötter, Perea, Aigner, Botnar, & Plank, 2009). At this stage, it is noteworthy to also consider two further properties of the NP:

2.1.2 The zeta potential

The Zeta (ξ) potential of a nanoparticle in a colloidal solution is the potential difference between the dispersion medium, i.e. the particle, and the stationary layer of fluid attached to the dispersed particle. It is therefore an indicator for stability in colloidal dispersions, as it shows the extent of repulsion between neighbouring particles bearing the same type of charge in a solution. A high zeta potential makes colloids of small size stabilized and resistant to aggregation. For the used PEI-Mag2 suspension in water, the zeta potential was determined as +55.0 ± 0.7 mV by Photon Correlation Spectroscopy (PCS) using a Zetasizer, which sufficiently stabilizes the particles against aggregation.

2.1.3 The Cluster Theory

As the magnetic properties and behavior of MNPs in the magnetic drug targeting applications are decisive aspects to successfully conduct a therapy, it is of great importance to understand the mechanisms exerting an influence on magnetically guided samples of particles. Magnetic interactions are therefore to be continuously evaluated before using the nanocarriers.

In the applications we are investigating in this work, mainly two types of interactions are interesting: the dipole-dipole interactions and the direct exchange interactions for touching particles (An-Hui Lu, 2007). As has been described in (Gleich, 2007), the overall magnetization of a number of MNPs in high concentration under the influence of an external magnetic field can be higher than the sum of their individual magnetic moments. This is explained by the Cluster theory. This means that loaded nanoparticles can exhibit the phenomenon that their ability to follow the magnetic field and their responsiveness to the external magnetic forces increase non linearly with the number of involved nanoparticles. This suggests nonlinearity in the resulting total magnetic properties of the samples.

The magnetic moment of an MNP can be written to

$$\vec{m} = \frac{m(\|\vec{B}\|)}{\|\vec{B}\|} \vec{B} \quad \text{with} \quad m(\|\vec{B}\|) = \iiint_V M(\|\vec{B}\|) \cdot dV \qquad (2)$$

where M is the magnetization of the MNP and is a function of the magnetic flux density in its nonlinear behavior

$$M(B) = M_s \cdot \left[\coth\left(\frac{\pi \cdot M_s \cdot B \cdot d_{equ}^3}{6 \cdot k \cdot T}\right) - \frac{6 \cdot k \cdot T}{\pi \cdot M_s \cdot B \cdot d_{equ}^3} \right] \quad (3)$$

(with M_s saturation magnetization, k Boltzmann constant, T temperature, B magnetic flux density and d_{equ} the diameter of the core of each particle)

Yet every magnetized particle generates a magnetic field that can be written to

$$\vec{B}(x,y,z) = \frac{\mu_0}{4\pi} \frac{3\vec{r}(\vec{\mu} \cdot \vec{r}) - \vec{\mu} r^2}{r^5} \quad (4)$$

(where $\vec{\mu}$ is the magnetic moment of the particle, $\mu_0 = 4\pi \times 10^{-7}$ V·s/(A·m) is the vacuum permeability and $\vec{r}$ is the vector pointing to the point in space in which the magnetic field is evaluated) and can contribute to the magnetization of the neighboring particles. Each particle would then experience a contribution B_j to its magnetization, generated by the neighboring n particles (assumed the system center is at the origin of the considered coordinates system and the z axis is pointing in the direction of the system's magnetic moment).

$$M_i = M_s \cdot \left[\coth\left(\frac{\pi \cdot M_s \cdot (B_0 + \sum_j^n B_j) \cdot d_{equ}^3}{6 \cdot k \cdot T}\right) - \frac{6 \cdot k \cdot T}{\pi \cdot M_s \cdot (B_0 + \sum_j^n B_j) \cdot d_{equ}^3} \right] \quad (5)$$

M_i is the resulting magnetization of a given nanoparticle surrounded by n neighbors.

To investigate this effect, a simplified measurement setup that precisely determines the total magnetic moment of different nanoparticle samples has been conceived, based on a high precision balance and a strong permanent magnet. The developed method allowed the measurement of the magnetic force on different nanoparticles' charges and shapes, and therefore the deduction of their total magnetic moment. Figure 6 shows the system setup as well as an exemplary nanoparticle suspension.

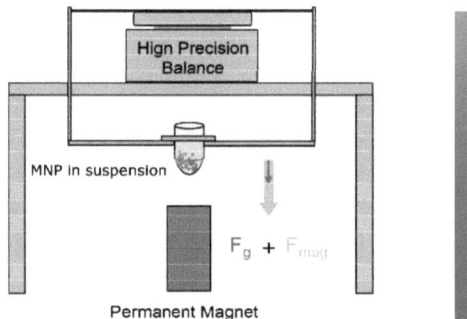 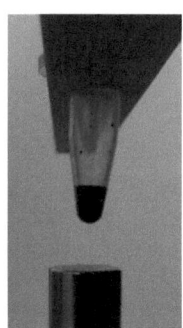

Figure 6: Measurement setup to determine the total magnetic moment of a given nanoparticle charge (left), Nanoparticle suspension exposed to magnetic force for magnetic moment measurement (right)

For a suspension of nanoparticles in an inhomogeneous magnetic field, at a point in space with a known gradient, the mass displayed on the high precision balance is directly proportional to the magnetic moment of the suspension

$$Displayed\ Mass - Weight\ of\ the\ sample = \frac{Magnetic\ Force}{Gravitational\ Acceleration}$$

As a direct consequence of the Cluster effect, a measurement of the cooperative behavior of a nanoparticle sample shows a nonlinear increase in its total magnetic moment with increased mass. Figure 7 depicts this effect for a sample of FeSi MNPs.

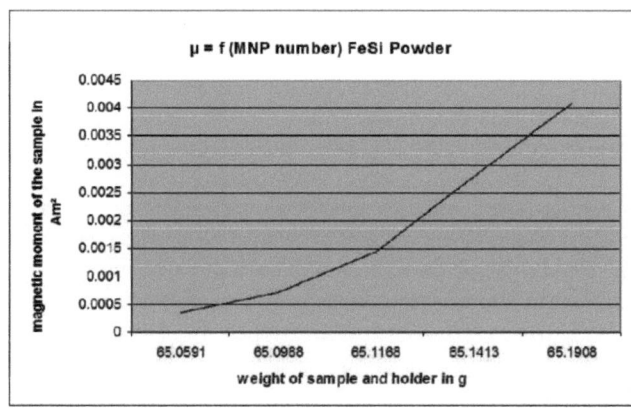

Figure 7: Nonlinear increase of the total magnetic moment of a sample with augmented sample mass as proof of the Cluster effect – here measured for FeSi magnetic nanoparticles

Now, having described these important aspects of particle-particle interactions, it is necessary to mention that in most cases, the effect of "Cluster Theory" is negligible, especially in constellations where the NPs are treated in a suspension:
- as the nanoparticles usually reach their saturation magnetization when fully exposed to the used magnetic fields (> 100 mT), for instance within a volume of 2x2x2 mm³ around the magnet tip. A particle within this area is fully magnetized and cannot benefit from the contribution of the neighboring particulates.
- as the zeta-forces stabilize the colloidal solution by sufficiently repelling the MNPs from each other.
For our calculations in this work, neglecting the magnetic particle-particle interactions is a worst case scenario (Gleich, 2007) as the real resulting magnetization of a given sample of MNPs is higher than the assumed one, given the fact of mutual magnetization when the applied magnetic field is below saturation level.
One further property that is often used to tune nanoparticles and optimize their behavior in an MDT environment is their shape.

2.1.4 The nanoparticle shape

The morphology of a nanoparticle is crucial for determining its behavior in industrial and especially medical applications. A change in the shape of a NP can cause an important change in its electromagnetic properties and with regards to biological interactions (Yuhong Mi, 2006). In particular, the magnetic properties of MNPs can be significantly tuned through shape arrangement. Furthermore, nanoparticles having modified designs –e.g. allowing them to exhibit a larger surface- are able to attach significant quantities of biological agents (T.K. Indira, 2010). Among others, researchers from the Johns Hopkins and the Northwestern universities have demonstrated in animal tests that a modification in a nanoparticle's shape could dramatically affect its efficacy in delivering gene therapy to cells (EurekAlert, 12 Oct 2012).
Although spherical nanoparticles are being used in most applications, there is a pertinent advantage in modifying this conventional shape to reach improvements in surface functionalization and environmental compatibility of the nanocarriers (Sang Jun Son, 2005).

Additionally to general observations made about nanoparticulates of non-spherical forms (Torchilin, 2006), the use and evaluation of spindle-type nanoparticles have also been reported in the literature (Yuhong Mi, 2006). To assess the usability of this particular shape in the experiments conducted in this study, nanoparticle samples (FluidMAG/SP-D 100nm) manufactured by the company Chemicell have been provided in a size range of 100 nm and exhibiting an elongated spindle shape.
The first attempt to dilute the MNPs in a HMDS (Hexamethyldisilazan) Primer

solution to facilitate its binding to the wafer surface resulted in particle agglomeration that could not be visualized through microscopy. A further attempt involved the dilution of the fluidMAG/SP-D particulates in alcohol (pure isopropanol). A sample of the obtained solution was then spin coated on a silicon wafer covered through sputtering by a 3 nm titan layer and a 2 nm layer of gold. The sputtering processes were conducted during 9 s each.

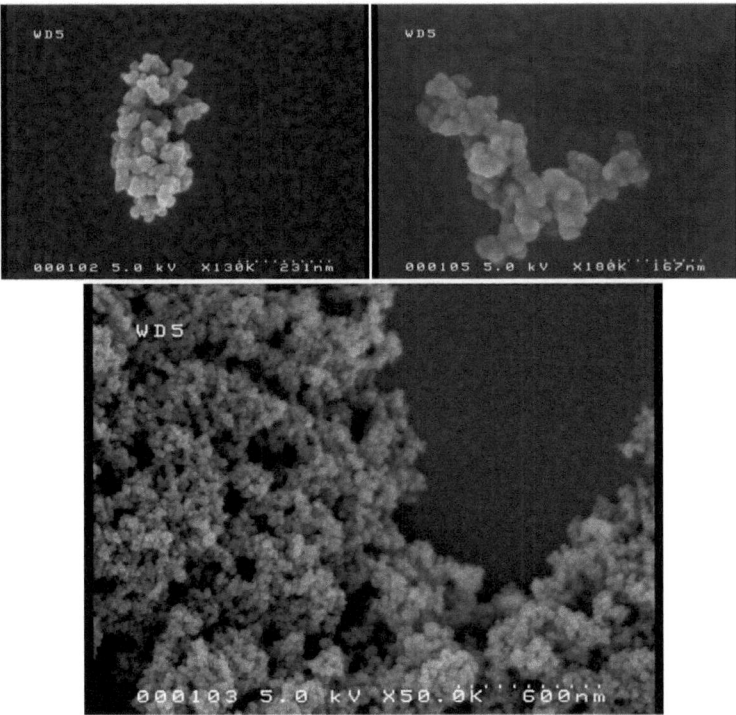

Figure 8: SEM images of spindle shaped nanoparticles after spin coating on a silicon wafer and sputtering (titan + gold) - top-left: a fluidMAG/SP-D particle/ top-right: two adjacent fluidMAG/SP-D particles/ bottom: the sample comprising multiple nanoparticles whereas their particular shape was lost through processing for SEM imaging.

The nanoparticles were then visualized using a Scanning Electron Microscope (SEM). Figure 8 depicts the images obtained through SEM and shows the loss of shape due to the processing of the sample through spin coating and sputtering.
This observed low stability against mechanical stress, added to the special conditions and techniques required for the synthesis of these particles, have been

the major arguments that led to the use of spherical magnetic nanoparticles in the further experiments of this research work.

2.2 Toxicity Vs. biocompatibility

Since the first uses of magnetic nanoparticles in the medical field, researchers have been evaluating and addressing their safety aspects. Interestingly, the small size of the nanomaterials, which is the main advantage leading to their successful integration into healthcare applications, is at the same time the major source of uncertainties with regards to the unintended effects of their use. In drug delivery for instance, nanoparticles present an enormous potential because they can be driven within the body to organs and tissues that usually aren't accessible through classical, bigger carriers. Yet, exactly this ability might represent a threat to a patient's health if the nanoparticles cannot be controlled in the organism. This is described as the toxicity of nanocarriers (Jain, 2008).
This toxicity can be related to several factors (Arruebo Manuel, 2007):

- chemical composition
- dose
- size
- shape/structure
- coating (surface chemistry)
- ability/tendency to aggregate
- biodegradability
- biodistribution
- solubility
- way of administration

Toxic effects of particulates in the body, either introduced through implants or aerosolized solutions, have been intensively studied, but the results and the amount of evidence reached are still non exhaustive. Further investigations will be needed. The statement that can be made is that nanoparticles can be entirely safe, and can also exhibit forms of toxicity. This usually leads to the necessity of a risk-benefit evaluation and trade-off. Nevertheless, several properties contributing to the biocompatibility of nanoparticles have been identified and tuned in different studies.

As a matter of fact, the modification of the surface chemistry remains an essential tool to minimize toxicological effects –as it defines how the particle is going to interact with the surrounding tisuue-, followed by the surface-to-volume ratio that can be controlled to avoid unfavorable biological responses. Sizes of less than 50 nm may raise most of the concerns, as particles with this size can easily enter the cells.

But even this observation becomes relative when we consider magnetic nanoparticles, for which it has been reported that larger particulates show higher cytotoxicity than smaller ones, when administered in concentrations in the range of 20-100 mg/ml (Arruebo Manuel, 2007). At this point, it is noteworthy to recall that the cytotoxiciy of a given nanosystem reflects its toxic effects on single cells, usually measured by the determining the concentration of the said substance that leads to the death of 50 % of a cell culture within 48h (Jain, 2008).

The measures that need to be taken to prevent unintended toxicological processes during application range from a safe bionanomanufacturing technique to a stringent immune surveillance after injection (Marco A. Zarbin, 2010), endorsed by the conduction of an adequate number of tests before applying the nanoparticles to the used animal model or patient body. Tests generally assess acute and long term toxicity, as well as the potential to stimulate inflammatory reactions. Through these checks, drug delivery nanoparticles should prove their non-toxicity and non-immunogenicity. They also have to fulfill hematocompatibility (compatibility with blood and blood cells), given the fact that most of the nanoparticles are led to tumors in the body through the bloodstream.

Further safety aspects are directly related to the route of administration that plays a major role with regards to the fate of the nanoparticles in the body, during and after application.

For instance, intravenous administration presupposes that the used nanocarriers are not only nontoxic and non-immunogenic, but also small enough to prevent embolizing capillary ducts or any other occlusion of the vascular system (R Jurgons, 2006). A nanoparticle injected into the blood is quickly subjected to the process of opsonization which consists in covering it with plasma proteins (e.g. immunoglobulins). This way, the nanoparticle is exposed to the recognition by the RES (reticuloendothelial system) defense mechanism that involves highly phagocytic cells derived from the bone marrow. The RES enables mononuclear phagocytes or monocytes to clear and deliver the trapped nanoparticles to the liver (Kupffer cells), spleen, lymph nodes and bone marrow. The phagocytosis and clearance of the NPs is performed in 0.5 to 5 minutes (Challa S.S.R. Kumar, 2005), removing thus the nanocarriers from the circulation and preventing them from accessing the target site, usually tumor cells. Depending on their size, the NPs may be delivered to the bile and excreted in the feces or filtered by the kidneys and incorporated into the urine. A general observation suggests that smaller nanoparticles are rapidly removed by renal elimination whereas larger carriers are taken up by the liver, spleen and bone marrow. In Table 5, the different removal processes and cells involved therein are shown.

Particle characteristic	removal process	involved cells
Large	phagocytosis	macrophages or dendritic cells
Small	Endocytosis	B or T lymphocytes
(biodegradable) magnetic NPs	Pinocytosis	any cells

Table 4: Different removal processes of NPs (Arruebo Manuel, 2007)

In the special case of magnetic particles, sizes of up to 4 µm lead to an elimination through the cells of the RES, mainly in liver (60-90 %) and spleen (3-10 %), sizes of more than 200 nm to a filtering by the spleen, and sizes below 100 nm mainly to phagocytosis by liver cells. The latter observation can for instance be used for treating liver tumors and diseases. It has also been reported, that smaller particles exhibit longer plasma half-life-times, i.e. that they circulate for much longer in the blood. Figure 9 summarizes the different cases of distribution and the routes taken by NPs after their injection into the body.

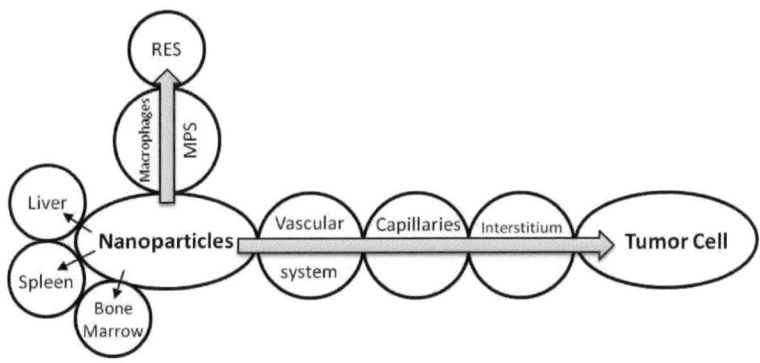

Figure 9: Fate of nanoparticles after injection for cancer therapy (Challa S.S.R. Kumar, 2005)

In general, it can be said that in order for a nanoparticle to accomplish its mission within the organism, it has to fulfill the key requirement of biodegradability or intact excretion. To ensure the latter in the case of nonbiodegradable cores, a specific coating can help avoid the exposure of the core until the excretion is assured by the kidneys. But even if the used material is biodegradable, an appropriate coating is needed to mask the NP from the clearance systems of the body so that it remains in blood circulation for a longer time.

When retained at target site, nanoparticles do not access the tumors through normal tissue, but use the hyperpermeable vasculature to enter and accumulate in the tumor interstitial region. There, the intrinsic lack of lymphatic clearance in tumoral tissue causes the retention of the nanocarriers, which is known as the EPR effect.

When biodegraded, the NP releases the drug load in the interstitium and the bioactive substance can then be absorbed into the tumor cells through diffusion.

As mentioned before, for Magnetic Drug Targeting applications, iron oxides with a core/shell structure are the most widely used nanoparticles. In particular, magnetite (Fe_3O_4) and maghemite (γ-Fe_2O_3) are preferred over other materials because of their biocompatibility and nontoxic character (Jana Chomoucka, 2010). Cobalt and Nickel do have better magnetic properties, yet they present a great hazard of toxicity and are therefore not suitable for human application.

Iron oxide is considered to be biologically safe, for it is easily degradable. When used in in-vivo applications, SPIONS are metabolized into elemental iron and oxygen by hydrolic enzymes. The iron is stocked in the normal body stores and is slowly embedded into hemoglobin through normal biochemical pathways known for iron metabolism. It has not been reported that iron oxide nanoparticles caused any toxicity effects such as alteration of the renal function, hepatic parameters, serum electrolytes or lactate dehydrogenase. The natural elevation of the iron level in the serum normally decreases after no more than 48 h without producing any symptoms.

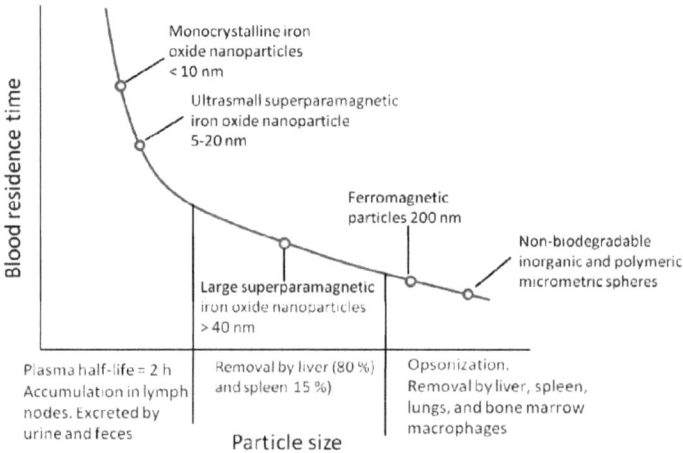

Figure 10: Dependence of the blood residence time of NPs on their size (Arruebo Manuel, 2007)

No side effects related to the use of iron oxide particles have been reported in the literature (Challa S.S.R. Kumar, 2005). This can be easily verified through the number of contrast agents based on magnetic Fe oxides available on the market (e.g. Feridex®, Resovist®) or the released iron oxide nanoparticles used as magnetic drug delivery systems (e.g. FluidMAG®, TargetMAG®). All satisfy the US regulations for use in human patients (Arruebo Manuel, 2007).

As already generally noticed for nanosized carriers, the time an iron oxide nanoparticle spends circulating in the blood also highly depends on its size. This is illustrated in Figure 10.

As implied above, there are still other aspects of toxicity to be considered in order to establish a fully safe therapy based on magnetic nanoparticles. In fact, toxicity can also be achieved by using iron oxide nanoparticles in an unbalanced concentration. In that case, the nanocarriers would represent no threat for cells as single particles but can as a group hinder the cells in their normal functioning through blocking its communication and interaction with its environment. Thus, a given nanoparticle type might be intrinsically nontoxic, but when applied in overdosed concentrations, it is susceptible to disturb the bioactivity of a targeted cell causing decrease in its vialbility.

Additionally, nanoparticles with particularly small sizes can cross the blood-brain-barrier (BBB) and access the inner compartments of the brain. An effect that can easily develop potential for malignancy. In a similar way, nanoparticles can be transferred from a mother to the fetus across the placenta and can constitute a danger for fetal cells.

On the other hand, it might be apparently controversial, but it is exactly the effect of toxicity that is needed to eliminate tumor cells in cancer therapy. Yet this has to be "controlled toxicity", i.e. the nanoparticles have to go through a safe voyage in the organism, land at target site and then deploy "useful toxic" effects in the therapy area.

Finally, if we extrapolate the notion of toxicity to cover all hazards of the MDT therapy, further aspects can also be considered, especially the magnetic field and its impact on organs, living tissue and bioactivity. In fact, the magnetic fields used in MDT have to be very large in order to produce significant effects. This suggests an assessment of the interaction of strong, static and dynamic magnetic fields with cells and cell communication and viability. Yet, sufficient clinical and safety experience has shown that the prudence and reserves towards magnetic fields reported in the 1980ies were misplaced and, for instance, the US Food and Drug Administration (FDA) revised the threshold set initially for allowed magnetic field values from 2 T in 1987 to 8 T (for adult patients) in 2003. No clinically relevant observation could be made about the capacity of large magnetic fields to reduce blood flow velocity or affect the behavior of erythrocytes containing micrograms of the Fe protein hemoglobin (Arruebo Manuel, 2007). Gradually, evidence is also being established regarding the use of dynamic magnetic fields in the low frequency range interesting for magnetic drug targeting applications. The frequencies implemented therein are surely far below critical values susceptible to generate hyperthermia effects or the uncontrolled triggering of action potentials in nerves (Dutz, 2008).

2.3 More efficient transport systems in MDT

Despite the superparamagnetic property of most of the nanoparticles used for magnetically guided drug delivery, it is still a great challenge to achieve sufficient retention at the target site when applying free particles. This is particularly due to the strong flow of the medium in which the drug carriers are navigated, be it blood or air. To increase magnetic responsiveness and therefore retention, magnetic nanoparticles can also be combined into larger conglomerates exhibiting a significant overall magnetic moment (Torchilin, 2006). Assuming that special care is paid to interparticle interactions, concentration and homogeneity of the resulting magnetic properties, nanoparticles are assembled to stable structures, either by conjugation into complexes comprising the bioactive load and the biocompatibility agents (Mykhaylyk, Steingötter, Perea, Aigner, Botnar, & Plank, 2009) or by aggregation into larger compounds (encapsulation). Figure 11 shows as examples for this approach a schematic view of a complex used in gene delivery and a conglomerate of single nanoparticles.

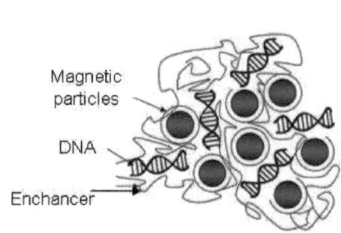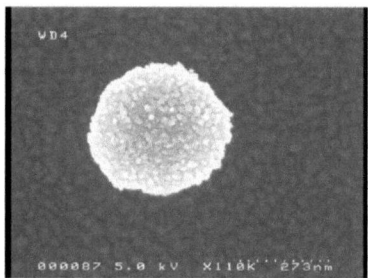

Figure 11: Left - Schematic illustration of an DNA-MNP-Enhancer complex comprising 110000 particles and 47000 plasmid copies as described and used in (Mykhaylyk, Steingötter, Perea, Aigner, Botnar, & Plank, 2009), Right – a conglomerate of magnetic nanoparticles (FluidMAG-D) from the company Chemicell® as visualized under scanning electron microscopy

Following the same principle of magnetic moment consolidation, the use of further sophisticated structures such as microbubbles has been reported. These consist in small gas filled cavities stabilized by a lipid, protein (albumin) or polymer layer. For applications in blood, the used gas has to exhibit low solubility. Microbubbles can therefore act as vehicles and carriers for magnetic nanoparticles –

particularly iron oxides- tightly bound or incorporated into their membrane, the nanoparticles being associated with different genes, siRNA, drugs or oligonucleotides. The obtained structures have sizes ranging from nano- to micrometers, and show favorable magnetic properties and a high binding capacity for nucleic acids.

This combination of microbubbles with their ability to enhance image contrast in ultrasonography and iron oxide nanoparticles with their high magnetic susceptibility and biocompatibility has already been exploited to create dual-modality contrast agents for both ultrasound and magnetic resonance (MR) imaging (April M. Chow, 2010) (Wen He, 2012). Coupling these two components helps overcome the issues of non-stability specific to microbubbles and the problems of non-specific penetration related to nanoparticles (Cai X., 2012). Using the same association in Magnetic Drug Targeting applications is therefore a logical evolution, and comes along with several advantages. For instance, encapsulating a great number of MNPs in the shell of a microbubble improves its magnetic properties and enhances its responsiveness to external magnetic fields, contributing thus to a better magnetic navigation and to a more precise targeting effect. This precision is enhanced when exploiting the fact that the membrane of a microbubble can be burst through exposure to ultrasound. By appropriately application, it can be achieved that only insonated target areas are affected by drug release and subjected to therapy.

A similar enhanced efficiency has also been reported by using magnetic lipospheres (Dialechti Vlaskou, 2010), a structure assembled through the shaking of a mixture of a cationic lipid, a nucleic acid, magnetic nanoparticles, aqueous buffer, and soybean oil. These lipospheres not only produce contrast in ultrasound images, they also exhibit higher magnetophoretic mobility resulting in an increased magnetic retention effect, in comparison with lose nanoparticles. Experiments conducted with these carriers led to functional DNA and siRNA delivery in vitro.

The same effect is also exploited when applying MDT for lung and airway therapy, where aerosols are enriched with great numbers of magnetic nanoparticles to create structures that exhibit higher magnetic responsiveness than single particulates. Further advantages of this combination are controlled delivery and release through magnetic guidance and activation, associated with the ability of performing CT and MR imaging (e.g. when using magnetite, gold nanoparticles or organic compounds) and the capacity to reach deep compartments of the lungs. Furthermore, the amount of drugs that can be loaded into a structure like aerosols is greatly increased when compared to single nanoparticles (Jain, 2008).

3 Magnetic guiding of nanoparticles for drug targeting in the blood vessels

In this chapter, the application of Magnetic Drug Targeting in blood vessels will be examined, starting with a model to describe the controlling variables and their interactions in the process of in-vivo magnetic guidance of nanoparticles, followed by a focus on the used magnet technology and leading to the experimental validation in animal models.

3.1 Modeling and Magnet

3.1.1 Modeling the MDT process

Several models have been reported in the literature to describe and eventually predict the mechanisms involved in magnetic drug targeting, in the blood as medium as well as in the airways. The common principle to all these models is the property of magnetic fields to exert forces on matter.

In the case of magnetic nanoparticle complexes, this force is described through the formula (1) introduced in chapter 2.1

$$\vec{F}_m = (\vec{M} \cdot \nabla)\vec{B}$$

In the microvasculature surrounding a target site, e.g. a malignant tumor areal, a nanoparticle carrying a therapeutic load is subjected to various forces ranging from the magnetic attraction, to the hydrodynamic and drag forces, over gravity and floatation or buoyancy. A schematic illustration of the most influential of these factors is depicted in Figure 12.

To describe the medium in which the nanoparticles are navigated, it is essential to have a close look at its components. Blood, for instance, consists of red blood cells, white blood cells, platelets and approximately 55 % of plasma. Red cells are the largest among the constituents and the most numerous, therefore prevailing in defining the mechanical properties of blood. In comparison, white cells and platelets have a negligible hemodynamic contribution (Leach, 2003).

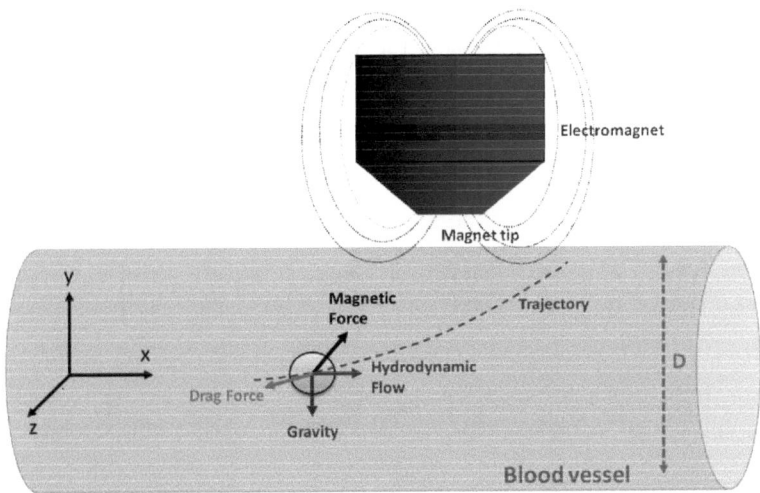

Figure 12: Schematic illustration of major forces acting on a magnetic nanoparticle navigated in blood vessels

At the normal shear rates observed in vessels, blood is assumed to be a homogeneous incompressible Newtonian fluid of viscosity $0.003 - 0.004$ kg/m·s at a temperature of 37°C. This means that its viscosity coefficient is independent of the velocity gradient. In large blood vessels, the flow is said to be a "plug flow" and exhibits more uniformity. In capillaries and smaller blood vessels, the flow has a rather parabolic front shape (Leach, 2003). Figure 13 shows both profiles.

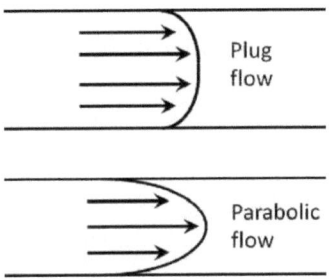

Figure 13: Front shapes of a plug flow (top) and a parabolic flow (bottom) in large and small blood vessels (Leach, 2003)

Considering the conditions and assumptions at hand, fluid motion in blood vessels can be described using the Navier-Stokes equations. These comprise three velocity equations and the continuity equation:

$$\rho\left[\frac{\partial u}{\partial t} + u\frac{\partial u}{\partial x} + v\frac{\partial u}{\partial y} + w\frac{\partial u}{\partial z}\right] = \rho g_x - \frac{\partial p}{\partial x} + \eta\left[\frac{\partial^2 u}{\partial x^2} + \frac{\partial^2 u}{\partial y^2} + \frac{\partial^2 u}{\partial z^2}\right] \quad (6)$$

$$\rho\left[\frac{\partial v}{\partial t} + u\frac{\partial v}{\partial x} + v\frac{\partial v}{\partial y} + w\frac{\partial v}{\partial z}\right] = \rho g_y - \frac{\partial p}{\partial y} + \eta\left[\frac{\partial^2 v}{\partial x^2} + \frac{\partial^2 v}{\partial y^2} + \frac{\partial^2 v}{\partial z^2}\right] \quad (7)$$

$$\rho\left[\frac{\partial w}{\partial t} + u\frac{\partial w}{\partial x} + v\frac{\partial w}{\partial y} + w\frac{\partial w}{\partial z}\right] = \rho g_z - \frac{\partial p}{\partial z} + \eta\left[\frac{\partial^2 w}{\partial x^2} + \frac{\partial^2 w}{\partial y^2} + \frac{\partial^2 w}{\partial z^2}\right] \quad (8)$$

$$\frac{\partial \rho u}{\partial x} + \frac{\partial \rho v}{\partial y} + \frac{\rho ww}{\partial z} = \nabla \cdot \rho \mathbf{u} = 0 \quad (9)$$

Where u, v, w are velocities in x, y, z direction respectively, ρ is the fluid density, η its viscosity, p is the pressure, g_x, g_y, g_z are the components of the gravitational force field in x, y, z. The variable t represents time and u is the velocity vector.

Solving this equation system goes through reducing it to time dependency:

$$\rho\frac{\partial \mathbf{u}}{\partial t} = \mathbf{F} + \eta\nabla^2\mathbf{u} - \rho(\mathbf{u}\cdot\nabla)\mathbf{u} - \nabla p \quad (10)$$

Here, **F** is a volume force field and represents other external body forces (M. Lohakan, 2007), for instance a magnetic force.

Using this equation, the hydrodynamic forces exerted on a nanoparticle submerged in an incompressible fluid can be determined (Thomas Weyh, 2004). Hereby, it is further assumed that for capillaries, hydrodynamic interactions between particles and vessels can be neglected, which is true when the diameter of the particle is much smaller than the diameter of the vessel. Additionally, steady flow conditions can be assumed as the calculated Reynolds number for the submerged particle in motion within the fluid is fairly small (Ovidiu Rotariu, 2005).

In fact, though the Reynolds numbers will be presented in detail in chapter 4.2, it can be said at this stage that for capillaries, where flow velocities are of the order of μm/s and the dimensions of the geometry below 100 μm, Reynolds numbers of less than one are often obtained.

This leads to the formulation of the hydrodynamic drag force acting on the nanocarrier as given by Stokes formula (Ovidiu Rotariu, 2005)

$$\vec{F}_{Stokes} = -3\pi\eta d_h \vec{v} \qquad (11)$$

Where η is again the viscosity of the medium (here blood), d_h the hydrodynamic diameter of the particle and $\vec{v}$ its velocity.

Considering the small size of the nanoparticles, we further assume that gravity and inertia (drag) forces are negligible compared to magnetic and hydrodynamic forces, so that a condition can be deduced to retain the nanoparticle and pull it to the magnet tip

$$\vec{F}_{magnetic} \geq \vec{F}_{hydrodynamic} \qquad (12)$$

Having (11) this condition can be translated into

$$\vec{F}_{magnetic} \geq \vec{F}_{Stokes} \qquad (13)$$

Assuming a reduced model of the problem in only two dimensions, all involved quantities can be described based on two unit vectors. The extrapolation to a 3-dimensional coordinate system is then easily implementable.

The Stokes force is re-written to

$$\vec{F}_{Stokes} = -3\pi\eta d_h \begin{pmatrix} v_x \\ v_y \end{pmatrix} \qquad (14)$$

And the parabolic profile characterizing the flow velocity in small vessels, e.g. capillaries, is given by

$$\vec{v}(y) = \frac{2\bar{v}}{R^2}(R^2 - y^2) \cdot \vec{e}_x \qquad (15)$$

where $\bar{v}$ is the mean velocity of the flow and R the radius of the vessel.

Assuming that the magnetic moment of the particle is aligned with the external magnetic field, this magnetic moment of one single nanocarrier can be re-written to

$$\vec{m} = \frac{m}{B}\vec{B} \qquad (16)$$

with $m = \|\vec{m}\|$ and $B = \|\vec{B}\|$

The magnetic force on a particle is therefore

$$\vec{F}_{magnetic} = (\vec{m} \cdot \nabla)\vec{B} = \left(\frac{m}{B}\vec{B} \cdot \nabla\right)\vec{B} = \frac{m}{B}(\vec{B} \cdot \nabla)\vec{B} \qquad (17)$$

$$\vec{F}_{magnetic} = \frac{m}{B}\begin{pmatrix} B_x\frac{\partial B_x}{\partial x} + B_y\frac{\partial B_x}{\partial y} \\ B_x\frac{\partial B_y}{\partial x} + B_y\frac{\partial B_y}{\partial y} \end{pmatrix} = \begin{pmatrix} F_{m,x} \\ F_{m,y} \end{pmatrix} \qquad (18)$$

When the particle is magnetically pulled within the flow towards the vessel wall, a magnetic force in the y-direction (perpendicular to the flow) is acting on it and is counterbalanced by a drag force governed by Stokes law.

$$F_{Stokes,y} = 3\pi\eta d_h v_y \qquad (19)$$

where η is the viscosity of the medium (here blood), d_h the hydrodynamic diameter of the particle and v_y the velocity of the particle in the y-direction.

With the condition (12), this leads in the worst case to

$$F_{m,y} = 3\pi\eta d_h v_y \qquad (20)$$

and thus to

$$v_y = \frac{F_{m,y}}{3\pi\eta d_h} \qquad (21)$$

The time needed for a particle to reach the vessel wall under these conditions and departing from the vessel centerline is (assuming v_y is not time-dependent)

$$t = \frac{R}{v_y} \qquad (22)$$

On the other hand, the particle is exposed to the magnetic field only in a constrained area surrounding the magnet tip for which a length L should be assumed. To cross this length, a particle flowing along the centerline of the vessel needs a time

$$t_{cross} = \frac{L}{\bar{v}} \qquad (23)$$

The time t needed to reach the vessel wall has to be less than t_{cross}

$$t \leq t_{cross} \qquad (24)$$

With (22), this leads to

$$v_y \geq \frac{R}{L} \bar{v} \qquad (25)$$

and thus to

$$F_{m,y} \geq 3\pi\eta d_h \frac{R}{L} \bar{v} \qquad (26)$$

Combining this result with (18) implies

$$\frac{m}{B}\left(B_x \frac{\partial B_y}{\partial x} + B_y \frac{\partial B_y}{\partial y}\right) \geq 3\pi\eta d_h \frac{R}{L} \bar{v} \qquad (27)$$

This condition on the needed magnetic field can be numerically computed, or –with the further assumption that the magnetic flux density is approximately fully defined by its y-component in the relevant area- transformed to

$$\frac{\partial B}{\partial y} \geq \frac{3\pi\eta d_h}{m}\frac{R}{L}\bar{v} \qquad (28)$$

Similarly, a condition on the magnetic field gradient in the x-direction can be obtained through considering reduced hydrodynamic forces acting on the particle when it reaches the vessel wall (Bernhard Gleich, 2007).

In practically all models, a quantitative evaluation is achieved only through numerical simulation, where the values of the governing quantities are computed using a combination of solvers for magnetostatics (involving the Finite Element Method) and computational fluid dynamics (CFD) (M. Lohakan, 2007) (Ushka K Veeramachaneni, 2007).

Based on calculations and experimental investigation performed by the research unit within which this thesis has been elaborated, the minimal value required for magnetic flux density is set to be above the saturation magnetization of the nanoparticles used in the corresponding experiments, for instance higher than 200 mT, whereas the field gradient should reach values above 10 T/m (C. Alexiou, 2006). Other scientific papers suggest values between 200 and 700 mT for the magnetic flux density and between 8 T/m and 100 T/m for the field gradient (P.A. Voltairasa, 2002), to achieve a magnetic effect in deeper tissue, whereas further investigations of magnetic targeting in models led to significantly lower values for the necessary gradient ranging from 0.03 to 10 T/m (Shin-ichi Takeda F. M., 2006).

3.1.2 The magnet

To perform the in-vivo experiments intended in this part of the thesis, a strong electromagnet exhibiting specific features has been designed and constructed.

The design of the electromagnet was based on the specifications deduced in chapter 3.1, with the ability to produce the necessary magnetic field properties in an active volume of 2x2x2 cm^3, which covers the whole volume of a small tumor area in a laboratory mouse model including its microvasculature. The entire system was conceived in a way to be manageable and of small size.

Prior to construction, the electromagnet was simulated using the simulation software COMSOL Multiphysics to investigate the optimal shape and physical parameters assuring the availability of a minimum magnetic flux density of 200 mT and a field gradient of 10 T/m in the targeted volume.

To compute and plot the magnetic flux density around the system tip, the model of the electromagnet was implemented in 2D as well as in 3D. Since the problem is symmetric to the z-axis, it was adequate to use the 2D axial symmetric mode in order to reduce complexity and processing time. For geometry visualization and construction purposes, a 3D model has also been revolved out of the 2D outline.

The simulation was conducted based on the Maxwell equations, where Maxwell-Ampere's law for the magnetic field $\vec{H}$ (A/m) and the current density $\vec{J}$ (A/m²) applies:

$$\nabla \times \vec{H} = \vec{J} \qquad (29)$$

For the magnetic flux density $\vec{B}$, Gauss' law states that

$$\nabla \cdot \vec{B} = 0 \qquad (30)$$

With the constitutive relation between $\vec{B}$ and $\vec{H}$ reading

$$\vec{B} = \mu_0 \mu_r \vec{H} \qquad (31)$$

Where μ_0 is the magnetic permeability of vacuum (Vs/(Am)) and μ_r is the relative magnetic permeability of the iron yoke (dimensionless).

Introducing a magnetic vector potential $\vec{A}$ obeying

$$\vec{B} = \nabla \times \vec{A} \qquad \text{And} \qquad \nabla \cdot \vec{A} = 0 \qquad (32)$$

the governing equation for the Magnetostatics mode can be derived to

$$\nabla \times \left(\frac{1}{\mu_0 \mu_r} \nabla \times \vec{A} - \vec{M} \right) = \vec{J} \qquad (33)$$

thus, reducing the basic input parameters to the relative permeability of the yoke material and the external current density.

Through optimizing the simulation results, the parameters have been determined as listed in Table 5.

Parameter	Value
Diameter of the copper wire	d = 1.2 mm
Cross-section of the wire	A_l = 1.13 mm²
Average length of the winding	l_m = 34.56 cm
Number of windings	N = 3714
Length of the coil	l = 1283.56 m
Mass of the coil	m = 12.95 kg
External current density	J = 1.79e6 A/m²
Relative permeability	μ_r = 4·10³ (Iron)
Output voltage	U = 41.12 V
Output current	I = 2.04 A
Power loss	P = 83.71 W
Adiabatic heating	θ = 40.3 K

Table 5: Parameters for the coil simulation

The final design comprised an iron yoke and an optimized tip geometry, as well as a adequately dimensioned copper coil. The iron yoke was necessary to enforce the magnetic flux density. The magnet system was designed in a modular way allowing for adjustments of the air gap between tip and yoke basis within a range of 3 to 7 cm and for the replacement of the tip, enabling, thus, the adaptation of the electromagnet to different sets of experimentation. Figure 14 shows the geometry of the model implemented for the magnetic system.

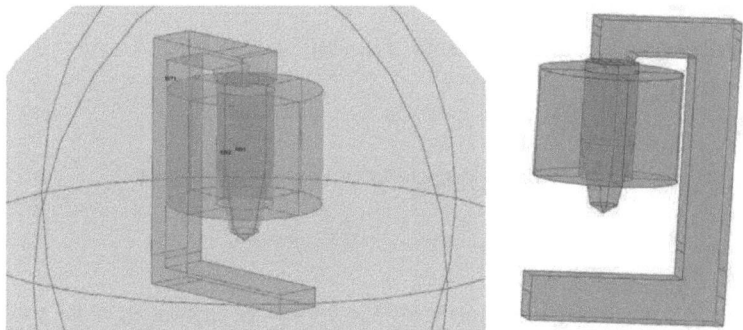

Figure 14: Model geometry of the designed magnet in 3D visualization

The simulation allowed the computation of the magnetic flux density in and surrounding the electromagnet, as well as the magnetic field gradient. Figure 15 shows the B-Field generated by the magnet system and the orientation of the field lines in the area around the tip.

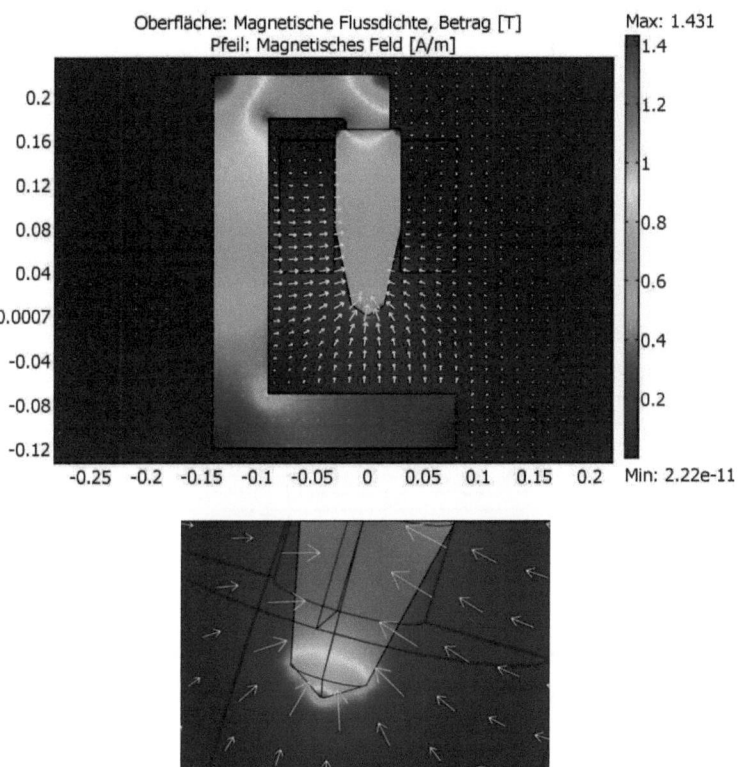

Figure 15: Simulated magnetic flux density of the designed electromagnet (top), and plot of arrows symbolizing the magnetic field lines surrounding the magnet tip (Hoke, 2008) (I. Hoke, 2008)

For a total current of 2.04 A applied to the coil, the magnetic flux density directly under the magnet tip was calculated to 588 mT. With the distance from the tip, the flux density falls then rapidly. The field gradient ranges from 27.08 T/m for z = 10 mm to 10.37 T/m at a distance of 20 mm from the magnet surface.

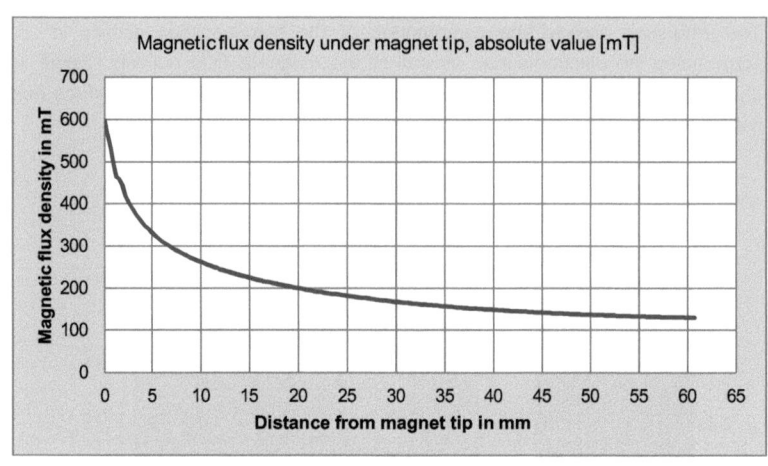

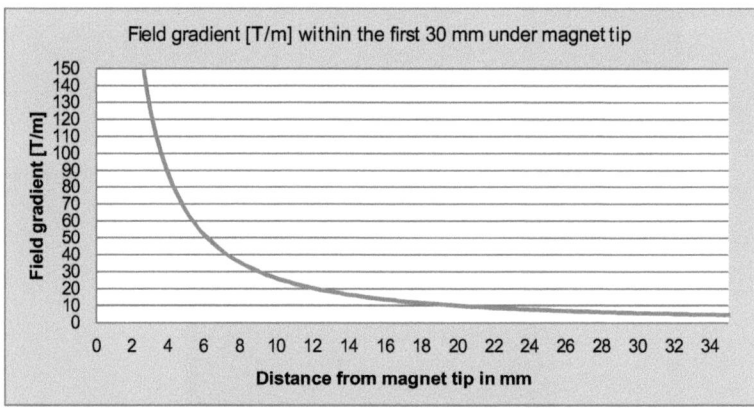

Figure 16: Plot of the magnetic field density and gradient with distance from the magnetic tip

After successful simulation, the electromagnetic system was constructed, based on the dimensioning plans displayed in Figure 17. The material used for all components, including c-shaped yoke and exchangeable tip, was iron.

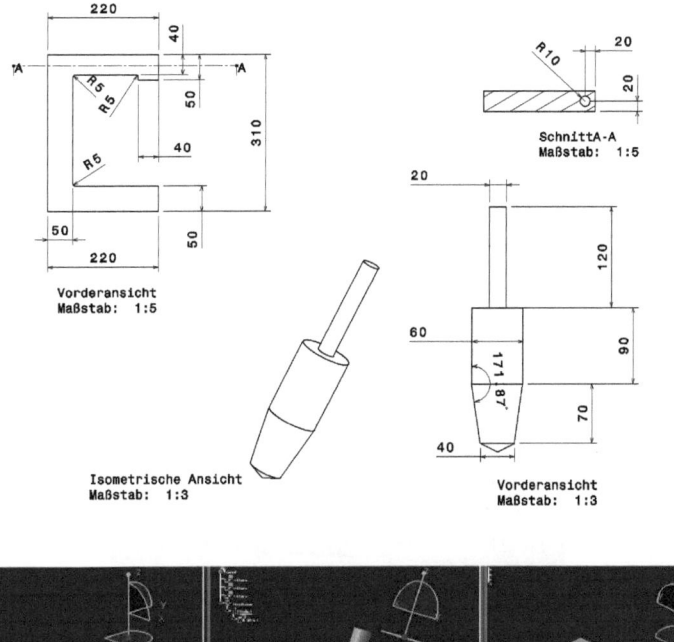
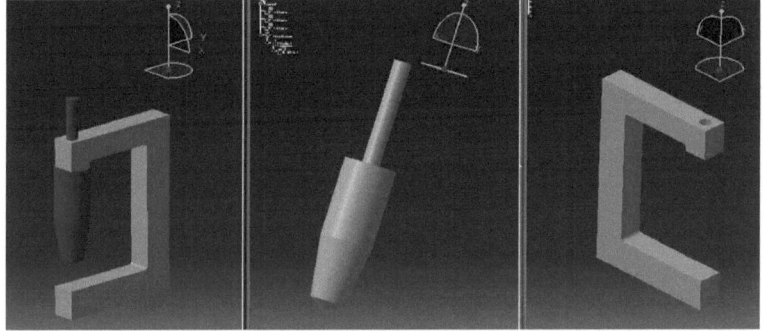

Figure 17: construction plans for the electromagnet, with a modular assembly design and an adaptable tip arrangement

The coil winding was performed using a 1.2 mm thick copper wire, and the variable air gap realized through a screw thread concept in the upper part of the yoke permitting the exchange but also an infinitely adjustment of the tip level. Figure 18 depicts the different components of the system as well as the assembled magnet.

Figure 18: Constructed experimental electromagnet, with its exchangeable tip system (bottom) and after assembly (top). The top left figure shows the magnet after galvanization and featuring a temperature measurement strip.

To prevent corrosion, the different components of the magnet were galvanized. Furthermore, 2 constant voltage generators (50 V/ 5 A) arranged in series were used to supply the system, delivering in total a voltage of 100 V and a current intensity of 5 A.

Finally, a thermal analysis was conducted through simulation showing an increase to 364 Kelvin (ca 90°C) in 2400 seconds (40 minutes) of the overall temperature of the winding, whereas the filling factor of 0.5 to 0.6 of the coil was not taken into account during calculation which can therefore be considered to be a worst case estimation of the thermal behavior of the system. To closely monitor the real time heating of the coil during experimentation, a reversible temperature measurement strip was mounted on the outer coil surface.

3.2 Animal experimentation and results

3.2.1 Targeting of microbubbles to the cutaneous blood vessels of a mouse (the in vivo application)

Whereas working with mathematical and simulation models or even cell lines is usually marked by a straight forward realization and reproducibility -including manageable difficulties during implementation-, experiments in vivo remain the most delicate part in research work, but also the most significant in terms of transferability to clinical application.

To explore and gather evidence about the efficiency of magnetic drug targeting, experimental work has been conducted in the scope of this thesis, involving pre-clinical trials on laboratory mice. This work was supported by the team of Prof. Dr. Krötz at his Physiology Institute, at the Medical Polyclinic of Munich.

The magnet system described in chapter 3.2 was used to realize microbubble retention in the blood vessels of the skin of a mouse. The choice of applying microbubble structures was based on first observations within the research activity that the retention of single, lose nanoparticles carrying oligonucleotides was not possible in the used model, therefore, a more efficient carrier structure was needed, as already explained in chapter 2.3.

The final goal of the experiments was to explore the site specific gene delivery in the vasculature of the skin through actively and selectively trapping magnetic nanoparticles carrying a therapeutic load at the target site.

To best expose the blood vessels in the specified area of the back skin of the mouse, a dorsal skin-fold chamber model was used, allowing additionally for visualization of the skin vasculature while keeping the animal alive, as shown in Figure 19. One major advantage of the chamber is that it can remain mounted on the mouse skin for several days.

Figure 19: The dorsal skin-fold chamber model, where part of the dorsal skin of a laboratory mouse is stretched along the window of a fixing device, allowing the visualization of the vasculature as well as the exposure to strong magnetic fields

The fixation chamber further fulfills the requirement that the targeted blood vessels are fully participating in the circulation. Interestingly, the used cylindrical Plexiglas cage does not impair or compromise the vital functions or calmness of the mice, as they are usually used to living in small structures and tight environments.

Animal selection:

Although male mice are more robust and likely to survive after the different manipulations needed to mount the skin-fold chamber, they present specific drawbacks consisting mainly in the biting scars they bear on their dorsal skin due to territorial fights with their male rivals as well as the strong pigmentation of their skin significantly compromising the visibility of the blood vessels in the chamber device. Therefore, female mice might be more appropriate for the presented investigation. However, it is necessary to reach a statistically correct observation and validation of the MDT effect, which imposes considering both genders in equal proportions for these experiments.
8 males and 8 females were involved in the study.

Animal preparation:

To prepare the animals for the MDT application, they are treated under narcosis one day before exposure, by implanting a micro-catheter (seen in yellow in Figure 17) into their carotid artery in order to have access to the blood circulation. This manipulation is highly delicate, as the dimensions of the blood vessels are exceptionally small and the catheter has to be accordingly shaped and further shrunk in size through heating. For the number of mice prepared, only 50% of them survived this intervention and participated in the study.

In a first step of the experimentation, only the trapping effect was investigated. For this, nanoparticles were coated with fluorescent oligonucleotides and incorporated into lipid bearing microbubbles that were injected intra-arterially using the already placed micro-catheters.

Figure 20: Flipped arrangement of the magnetic system with the upwardly oriented tip, allowing for better accessibility to the vessels of the skin-fold chamber of the mouse

The layout of the magnetic system was transformed based on its modular design, by reversing the direction of the magnet tip and flipping it on top of the iron yoke, as displayed in Figure 20.

This arrangement allowed an optimized placement of the magnet tip under the dorsal skin-fold chamber (starting from a distance of 1 mm), as shown in Figure 21 and therefore a highly increased exposure to the magnetic field.

Figure 21: Fixation device including skin-fold chamber mounted on top of the magnet tip

The solution featuring the microbubbles loaded with the PEI-Mag2 nanoparticles as well as a GFP-enriched viral shell containing no DNA material was then injected into the catheter and brought into circulation with the blood of the mouse.

Through this injection (150 to 200 µl), the total blood volume ranging from 1.5 to 2 ml and flowing in the whole vasculature of the 20-30 g animal is dramatically augmented (~ 10 %). This is comparable with adding 750 to 1000 ml of fluid to the human blood volume, comprising normally 7 l, within only 20 minutes. Therefore, it was not feasible to subject the mice to a second injection during one experiment and the effect was investigated in one single application.

To keep the mouse under control and therefore avoid artifacts in experimental results or images, narcotizing substances were used, yet without altering the blood flow, be it with a negligible deceleration of the cardiac pulse.

3.2.2 Results:

For a qualitative assessment of the targeting efficiency, optical verification under intravital microscopy was assured by placing the entire dorsal skin-fold chamber under an Axiotech Vario microscope (Zeiss, Göttingen, Germany) and visualizing the fluorescence expressing sites with a digital camera mounted on the microscope (HDcam Hamamatsu 1394 C8484-05G, Hamamatsu, Germany), as depicted in Figure 22. No interference between the magnetic and the imaging systems was reported, as the main components of the object lens compartment are amagnetic.

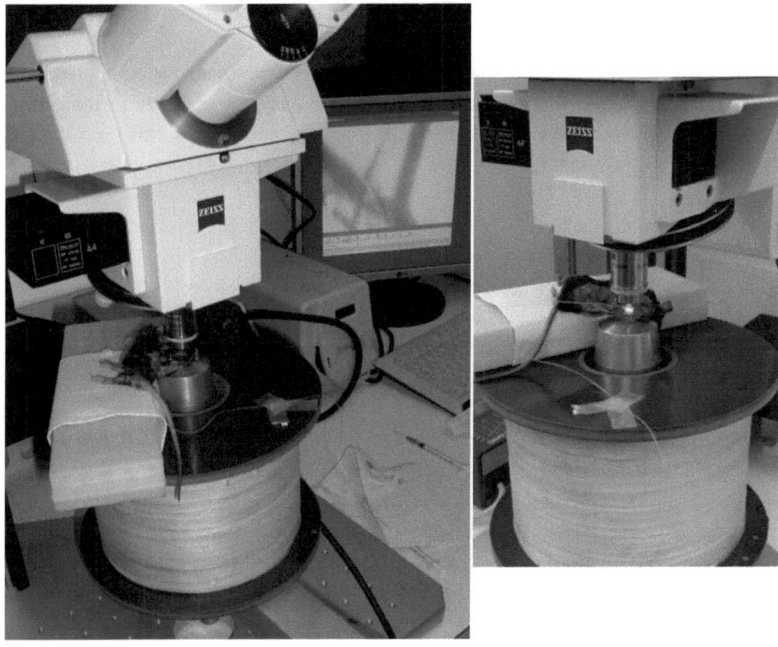

Figure 22: Zeiss Microscope placed few millimeters above the skin-fold chamber that exposes the confined part of the dorsal vasculature to a high magnetic field (1039 mT). The configuration features fluorescence imaging (right) and enables real time assessment of the retention efficiency on a display (left)

Magnetic trapping of Microbubbles in the microvasculature of the dorsal skin:

After injection of the lipid microbubble suspension, the electromagnet was activated and supplied with a voltage of 100 V leading to a current intensity of 5 A. The magnetic flux density at the tip of the magnet was measured to 1039 mT.

Using microscopy, the vessels of the dorsal skin of the mouse were optically inspected for appearance of microemboli or thrombi after magnet activation. The observation was made that immediately after magnet field triggering, the blood flow started slowing down progressively until it reached cessation in the visible capillaries and arterioles of the skin chamber after few minutes. This observation is termed Microembolization and certifies the creation of a thrombus due to the agglomeration of nanoparticle carrying microbubbles at target site.

The magnetic attraction was so strong and the exposure of the microbubbles in the vessels so high that a retention effect was evidenced already in the artery, despite blood flow velocities reaching 20 cm/s therein. This retention could be – reproducibly- generated.

Whereas microbubbles tend to naturally embolize smallest capillaries of 5 to 7 μm just due to their size (ranging through aggregation capabilities from 1 to 12 μm), the embolization of bigger arterioles of up to 500 μm, as observed in this experiment, was only reached by magnetic retention.
After 4-5 minutes, a further observation was made that the microbubbles started gathering in the veins and retention was achieved in both blood vessel types.

Immediately after magnet deactivation, the microbubble agglomeration was slowly dissolved and normal blood flow gradually re-established.

It is at this stage noteworthy to recall that the observed thrombi represent no threat to the organism of the animal, as they are rapidly dispersed after retention stop and due to the significant distance separating them from the heart, lungs and brain.
Figure 23 shows the fluorescence tracking images obtained and the effect observed in the arteries during magnetic trapping.

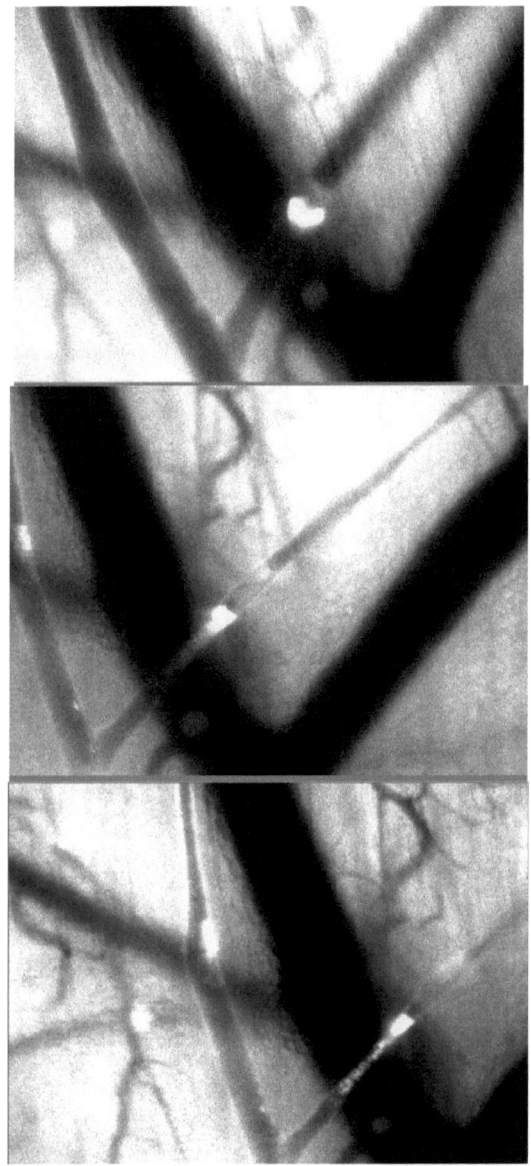

Figure 23: Fluorescence tracking under microscopy of microbubble agglomerations in the arteries of a mouse (dorsal skin). From top to bottom, the pictures show the gradual formation of the agglomeration, including at bifurcation, under magnetic field influence.

The retention effect can be tuned by changing several parameters of the experiment, yet it can be optimized by adjusting the concentration of microbubbles used.

The successful retention leads to the deployment of the therapeutic agent at target site, enabling thus the proliferation within 48h of new capillaries in the case of angiogenesis supporting biosubstances. Similarly, the therapy can also aim at the opposite, namely blocking the formation of new blood supply, by using specific oligonucleotides. The inhibition of pathological angiogenesis or the overexpression of a defined molecule is one of the main applications of magnetically targeted gene delivery.

When inspecting the genome, mice are considered to be close enough to human beings (Chris Gunter, 2002), to enable transposing several results to pre-clinical application, especially in the case of new blood vessel creation. This suggests that reaching comparable magnetic trapping results of nanoparticles in human vasculature is plausible.

As a continuation of these experiments, further tests involving quantitative methods and ultrasound destruction of the microbubbles were presented in the paper of Mannell et al. (Hanna Mannell, 2012), where gene delivery to specific target sites in the vasculature of the mouse has been achieved based on a systemic application, enabled through magnetic targeting of nanoparticle coated microbubbles that were disrupted by ultrasonic waves. The microbubbles loaded with dsRed plasmid DNA were injected intra-arterially and retained at site using the magnetic system presented in chapter 3.2. By quantitative real-time PCR, dsRed expression was detected only in the targeted area, and when both magnetic field and ultrasound destruction were involved.

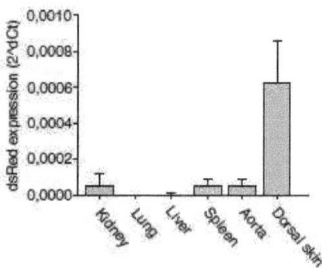

Figure 24: Results of site specificity of gene delivery mediated by magnetic microbubbles as achieved in the skin-fold chamber vasculature of a mouse (Hanna Mannell, 2012). The expression of the applied plasmid DNA was detected in a significantly differentiated way in the targeted site.

Here, it should be pointed out that the application of ultrasound waves after retention has to be limited to a short time in order to prevent the over-permeabilization and therefore deterioration of the targeted blood vessels

4 Magnetically guided nanoparticles for drug targeting in the lungs

Drug Targeting basically involves three major steps: the active guiding and steering of carriers loaded with therapeutic substances, the selective overcoming of natural barriers within the organism such as the blood-brain-barrier or cellular membranes, and setting free the active load at target site, so that a therapeutic effect can be achieved. The targeting can be significantly enhanced through the use of magnetic nanoparticles as carriers and the application of adapted magnetic fields to control their navigation. This mechanism can be implemented not only in the blood vessels, but it is also suitable for other media circulating in the body such as the air.

In fact, targeted delivery of magnetic nanoparticles, accordingly encapsulated in aerosols, can be applied to treat diseases in the lungs, which is generally defined as the technique of Lung Drug Targeting (LDT).

Especially for cancer therapy in the lungs, but also for other ailments of the respiratory system, LDT presents an important potential, given the fact that the internal tissues of the pulmonary pathways are intrinsically more difficult to access through blood than through air circulation.

The implementation of this technique involves nebulizing a drug solution and mixing it with magnetic nanoparticles to generate aerosol droplets exhibiting magnetic properties. These „nanomagnetosols" can be steered through external magnetic fields, and retained and concentrated at target sites (Dames, 2007), which would increase the locally delivered active dose compared to classical aerosol therapy. This approach is most promising in relation with the application of chemotherapeutic drugs to lung carcinomas, where a focused deployment of the therapeutic effect automatically leads to the reduction of side effects generally observed in the whole organism due to the non-selectivity of the used drug or its application mechanism.

Gradually, significant advances in the exploration of Lung Drug Targeting have been reported. Since the in-vitro experimentation of Ally et al in 2005 who used a model to study and prove the feasibility of targeted delivery of chemotherapeutic substances attached to inhaled aerosols (Javed Ally, 2005), a remarkable progress has been achieved by Dames et al in 2007, when the successful pre-clinical application of this method was demonstrated in mice lungs (Dames, 2007).

How far these results can be transposed to human treatment is still a subject of investigation. Hereby, it is of great utility to take into consideration the fact that a diseased lung is usually significantly different from a healthy one. As a matter of fact, pathologies in certain areas of the lung inevitably lead to a degeneration of their function. Beyond this constraint, the lungs are also subjected to a continuous motion due to breathing, which imposes an additional challenge on the treatment methodology and the planning of the magnet application (Gläßner, 2008) (Dahmani Ch., 2008). A further important aspect that needs to be investigated and strongly considered in the scope of LDT is the mucociliary clearance of particles deposited in the tracheobronchiol region. For instance, studies reported that only 5 % of the particles having a diameter of 6.5 μm that are retained on the inner walls of the upper pulmonary pathways escape from mucociliary clearance.

These are but some of the most pertinent factors that have a significant impact on the success of a Lung Drug Targeting procedure.

In this chapter, an investigation of a number of selected, reportedly crucial aspects of this technique will be conducted, based mainly on simulations and experimental validation.

To start, the main requisite for the successful implementation of this therapy is a sound understanding of the anatomical constraints related to malignancies of the lungs and the upper pulmonary airways.

4.1 Physiology of the human lungs

The lung is the main respiration organ and is responsible for the transport of oxygen from the atmosphere into the bloodstream, and the release of carbon dioxide from the bloodstream into the atmosphere. The anatomy of the lungs is best understood when the passage of air through the outside to the alveoli is followed and described: after progressing through the mouth or nose, air passes in the pharynx or throat on its way to the trachea (windpipe). The trachea is divided into two main airways upon reaching the lungs, these are the bronchi. One bronchus serves the right lung and the other the left. The bronchi further subdivide into increasingly smaller branches called bronchioles. Bronchi and bronchioles form the so called bronchial tree. The divisions of this tree are denoted generations, and at the level of the 23rd generation, the bronchioles run into alveolar ducts that conduct the air to clusters of small air sacs called alveoli (Ozer, 2007). The main constituents of the bronchial tree are further illustrated in Figure 25.

Observed from the outside, human lungs are located in two cavities on either side of the heart. The right and the left lungs are not identical and are subdivided by fissures into lobes, with three lobes on the right and two on the left lung. These lobes are further divided into segments and then into small, hexagonal compartments called lobules. Within the lobules lay the alveolar structures responsible for the transfer of the oxygen from the air transported through the respiratory system into the bloodstream.

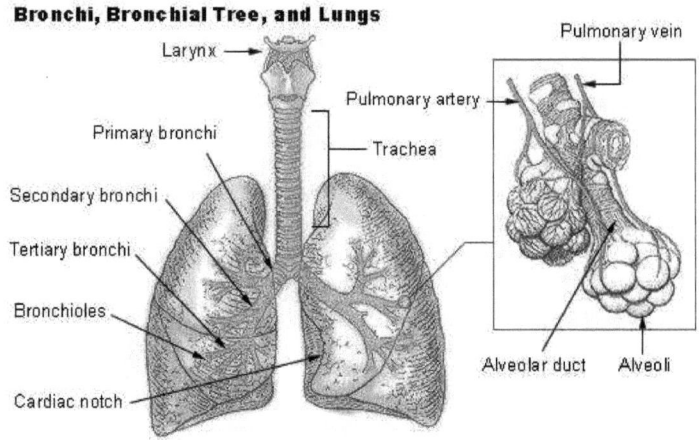

Figure 25: Illustration of the main constituents of the human lung comprising the trachea, bronchi, bronchial tree and ending up at the alveolar level (National Cancer Institute, 2012)

To understand the challenges related to applying Magnetic Drug Targeting to the respiration system, it is of great utility to further observe the involved lung substructures. For instance, a basic treatment scenario would suppose the patient inhaling a nebulized solution containing airborne magnetic nanoparticles loaded with therapeutic substances and encapsulated into aerosols. These magnetic aerosols, also described as nanomagnetosols (Dames, 2007), have to pass the upper airways and reach the targeted sites deep in the bronchial tree. Among these upper pulmonary pathways, the trachea, main bronchi, and approximately the first dozen divisions of smaller bronchi carry rings or patches of cartilage in their walls that prevent them from collapsing or blocking the flow of air. The rest of the bronchioles and the alveoli do not incorporate a cartilage and are therefore very elastic. This allows them to respond to pressure changes as the lungs expand and contract (Kübler, 2007).

The thoracic muscles enable the deformation of the thorax space, and can generate a negative pressure at the lung periphery, which increases the volume in the lungs through air inspiration. The expiration is in the contrary achieved through the creation of a positive pressure that pushes the air out of the lungs and pulmonary pathways. In the case of artificial, mechanical ventilation, these pressure conditions are different.

Given the fact that the right lung has a bigger volume than the left one, there is a tendency to suppose that inhaled airborne substances "prefer" flowing into the right lung lobes.

In the case of pulmonary drug delivery, the aspects related to particle deposition become of utmost importance. For instance, the size of the inhaled particles is decisive for their fate within the pulmonary pathways. This size is defined through the aerodynamic diameter which is determined by the actual size of the particle, its shape and its density (Ozer, 2007). For an aerodynamic size between 0.5 and 3.5 μm, particles bypass the bronchial airways in the inspiration phase and reach the deep compartments of the lungs. If the size range is between 3.5 and 6.0 μm, particles are able to penetrate -to different extents- beyond the central airways and reach the peripheral area of the lung, provided the inspiration is conducted at a low flow rate. Particles with a larger aerodynamic size are subjected to their inertial mass and therefore tend to impact already in the upper pulmonary pathways, whereas smaller particles, with diameters below 0.5 μm, are more exposed to thermal interactions with the air molecules surrounding them, which causes them to diffuse to the surfaces in the respiratory tract during the inhalation phase (Ozer, 2007).

Foreign particles that touch the inner walls of the respiratory tree get stuck on a thin fluid film of mucus covering hair-shaped structures called the cilia. These perform a coordinated movement within the mucus layer, that propels up the trapped particles to the pharynx and mouth, where they are swallowed (Gläßner, 2008). This process termed mucociliary clearance is a natural host defense mechanism of the airways and has a significant impact on the Lung Drug Targeting procedure, as every trapped particle or nano-carrier, including its freight, is considered to be lost and does not contribute to the therapeutic effect.
A good planning and implementation of the LDT technique should therefore involve solutions for the mucociliary clearance issue, and reduce, if not avoid undesired contact between the airborne particles and the inner walls of the respiratory pathways.

Further challenges of Lung Drug Targeting are intrinsically related to the parameters of respiration, such as temperature, degree of humidity in the inhaled air, the position of the patient and the lungs during treatment, the use of artificial ventilation under patient anesthesia, the air flow rate, etc. For instance, the introduction of catheters or tubes in the airways during application causes inevitable distortions not only in the airway geometry, but also in the flow dynamics and particle behavior. This for example makes it almost unfeasible to correctly assess the velocity distribution of air and particles in the pulmonary pathways when measurement catheters are used (H. K. Chang, 1982).

4.2 Simulating the Lung Drug Targeting procedure

In the last years, computer simulation has constantly gained in importance as a strong and non-invasive tool to study airflow and particle transport and behavior during respiration, the main application being aerosol therapy for pharmacological purposes.
To precisely and faithfully reflect in vivo conditions, simulations have to accurately reproduce the morphology of the respiratory system, as well as the breathing conditions. This implies a physiologically realistic three-dimensional modeling of the lungs and the upper airways (Ted B. Martonen, 2001).
In this chapter, an analysis of the behavior of airborne nanoparticle-loaded aerosols in the upper pulmonary pathways is conducted in a simplified lung model. The computation of the fluid dynamics and particle trajectories is mainly realized with the CFD software ANSYS Fluent (based on the Finite-Volume-Simulation and analysis program for numerical fluid mechanics problem solving) in combination with the Magnetostatics Module of the FEM based software package COMSOL Multiphysics.

For the simulation of fluid dynamics, numerical methods and algorithms are used to describe the motion of fluids, under a number of assumptions. For instance, the studied fluid is considered to be a Newtonian fluid, with a constant density and a constant viscosity.

This is translated into a set of nonlinear coupled partial differential equations that has to be solved. These equations represent the laws of conservation of mass (continuity), momentum, and energy, corresponding to the Navier–Stokes equations in a modified form (zero viscosity and heat conduction terms) (Issa, 2010).

For the simulation undertaken and presented in this work, the temperature of the inhaled air has been assumed constant, so the first law of thermodynamics has not been taken into consideration and the solution of the system relied on the set of equations (6), (7), (8) and (9) described in chapter 3.1., comprising the Navier-Stokes equations and the continuity equation. In an attempt to reduce complexity, geometries of tubes have been reported in the literature as adequate models to describe respiratory pathways. In these models, the flow of a fluid is said to be laminar if the average velocity $\overline{U}$ is sufficiently small, whereas for higher velocities, turbulences and eddies occur, and the flow is said to be turbulent.
At this stage, it is important to point out a number of relevant quantities that can be used to describe the fluid mechanics problem in general:

- the entrance length L_E: is the length from the tube entrance necessary for

complete hydrodynamic development and establishment of flow profile.
- the (dimensionless) Reynolds number Re: is used to characterize different flow regimes, such as laminar or turbulent flow. For instance, low Reynolds numbers indicate a laminar flow, in which viscous forces are dominant, and the fluid motion is smooth and constant, whereas high Reynolds numbers characterize a turbulent flow dominated by inertial forces, generating chaotic eddies, vortices and other flow instabilities.
The Reynolds number is given by

$$Re = \frac{\rho \bar{U} L}{\eta} = \frac{\bar{U} L}{v} \qquad (34)$$

where ρ is the density of the fluid, η its dynamic viscosity, L a characteristic travelled length of the fluid, and v the mean velocity of the object relative to the fluid in m/s.

For Reynolds numbers below a critical value $Re_{critical} \approx 2300$, the motion of the fluid in a tube model can be described as laminar. Above this value, the flow is said to be instationary (Gläßner, 2008).

For laminar profiles, the following relation can be assumed for Re and L_E

$$\left(\frac{L_E}{2R \cdot Re}\right) \in [0.0287m;\ 0.0575m] \qquad (35)$$

where R is the radius of the observed tube (Gläßner, 2008).

- the (dimensionless) Stokes number: describes the behavior of particles suspended in a fluid flow. It is defined as the ratio of the resistance force (stopping distance of a particle) to friction (a characteristic dimension of the obstacle).

$$St = \left(\frac{\rho_P d_P^2}{18\eta}\right) / \left(\frac{2R}{\bar{U}}\right) \qquad (36)$$

$$= \frac{\rho_P d_P^2 \bar{U}}{36\eta R} \qquad (37)$$

$$= \frac{\rho_P d_P^2 \dot{V}}{36\pi\eta R^3} \qquad (38)$$

where ρ_P is the aerosol particle density, d_P is the diameter of the aerosol, R is the radius of the tube, and $\dot{V}$ is the air flow rate.

The Stokes number is used in the literature to measure to which extent a fluid-borne particulate follows the flow. In the case of Lung Drug Targeting, this prediction ability is altered due to the impact of the applied magnetic field on the trajectories of the nanoparticle-loaded aerosols (Issa, 2010).

In this work, the magnetic field considerations have been conducted for a multiparametric analysis of the particle deposition behavior in the respiratory tract under magnetic force application, focusing mainly on the following parameters (Issa, 2010):

Varied quantity (parameters):	Values:
Flow rate $\dot{V}$ [l/min]	15, 20, 25, 30
Aerosol diameter d_p [μm]	1, 2.5, 3.5, 5, 10
Magnetic moment of a whole aerosol [Am²]	$7.35 \cdot 10^{-12}$, $6.46 \cdot 10^{-14}$, $1.102 \cdot 10^{-11}$
Angulation of magnetic tip from vertical reference [°]	0, 45, 60, 90
Activation time of magnetic field [s]	0, 0.05, 0.1, 0.2, 0.3, 0.4, 0.5

Table 6: Parameters varied to study particle deposition under magnetic force application in a lung model geometry (Issa, 2010)

Beyond these quantities that have been varied in the scope of the study, a number of quantities have been set constant, for instance the air density $\rho = 1.109 \ kg/m^3$, and the air viscosity $\eta = 1.905e^{-5} \ kg/ms$, both independent from temperature. It is also assumed that the air inhaled during ventilation is conditioned at entrance to reach a body temperature of 37.5°C. For the used aerosols, a particle density of $\rho_P = 1005.25 \ kg$ is considered.

4.2.1 Implemented model geometry for the simulation

For a realistic asymmetric structure of the human lungs and respiratory system, the Model 1 presented by Horsfield et al. has been adopted (K. Horsfield, 1971). This model relies on the percental distribution of the outlet air flows, as detailed in Table 8.

The simulation being based on the scenario of intubation, the airway model had to exclude the extra-thoracic upper airways, i.e. the mouth cavity, the oropharynx and the larynx, and to focus on the intra-thoracic central airways of up to 4 generations from the zeroth generation (G0) of the trachea to the fourth generation (G4) of the airways, with a total of 10 peripheral small airways. Figure 26 depicts the specific lung airways in the conducting tree.

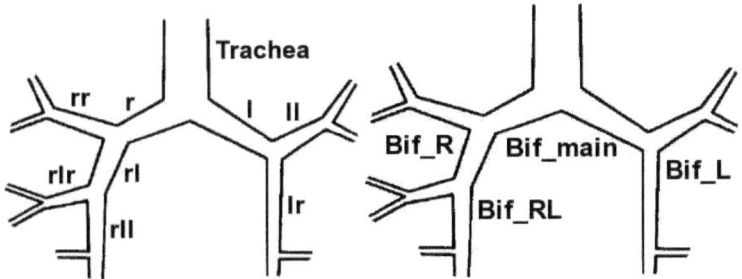

Figure 26: Denomination of the upper respiratory pathways and main bifurcations, as derived from Horsfield's Model 1 (K. Horsfield, 1971)

Airway segments are identified by a nomenclature comprising the letters "l" for left and "r" for right side.

Branch	Radius [mm]	Length [mm]	Flow [% trachea]	Bifurcation angle [°]	Curvature radius [x radius]
Trachea	8.00	100.00	100.0000		
L	6.00	50.00	45.0342	73	9.0
Ll	3.75	16.00	18.9193	48	7.0
Lr	4.00	11.00	26.1149	44	12.6
R	5.55	22.00	54.9658	35	6.0
Rr	3.65	15.60	18.9193	63	3.4
Rl	4.45	26.00	36.0465	15	4.6
Rlr	2.60	21.00	9.9316	61	16.0
Rll	3.20	8.00	26.1149	15	11.8

Table 7: Lung modeling data according to Horsfield's Model 1

Given these data, the analyzed simulation area was defined as beginning from the trachea and reaching the accesses of the five human upper pulmonary ducts. The geometry was created using the Physiologically Realistic Bifurcation (PRB) approach, in which fully realistic flow fields can be computed. Using this geometry, an unstructured polyhedral mesh has been created featuring 651,918 cells in the case of non-intubation and 716,000 cells when including the intubation. Figure 27 shows both geometry and mesh (Issa, 2010).

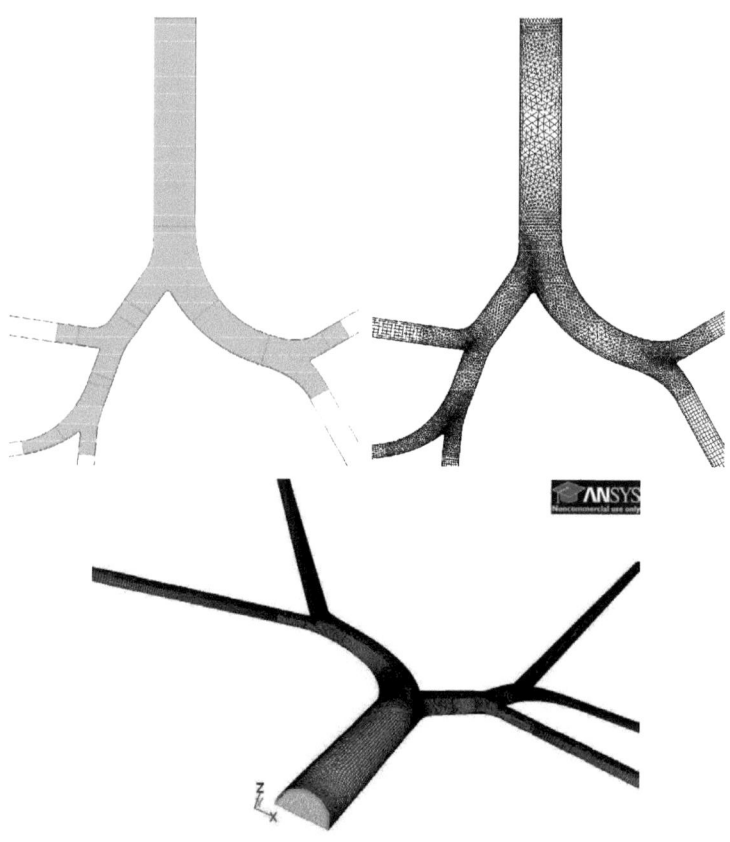

Figure 27: Designed geometry (top left) and mesh (top right and bottom) of the upper pulmonary airways, based on the Horsfield's lung data and the physiologically realistic bifurcation computation approach (Gläßner, 2008)

In this study, similarly to almost all studies reported in the literature, only the impact of a set of factors and parameters could be investigated. An exhaustive analysis of all variables is not feasible due to the enormously increasing problem complexity. Although the processing capabilities of the newest generations of computers has tremendously progressed over the last 20 years, a compromise still needed to be made to balance computation time, complexity and accuracy of results.

Nevertheless, indications derived from the literature are a good basis to identify and select the most significant simulation parameters, assumptions and models (Gläßner, 2008). Common assumptions that are reported in the literature have also been implemented into this study and include (Issa, 2010):

- a rigid geometry of the airways
- smooth inner walls of the respiratory pathways (no tracheal rings)
- no slip conditions at inner walls
- laminar air flow (in geometry with no intubation)
- stationary flow (steady state, no time variation of field quantities)
- discrete spherical particles as aerosols (with constant mass)
- unified aerosol diameter
- particle deposition at wall contact
- neglected forces acting on the aerosols stemming from Brownian motion and gravitation
- neglected interactions between aerosols and with the continuous phase (air)

4.2.2 Magnetic forces and particle trajectories

The deployment of Magnetic Drug Targeting in the lungs necessitates the retention of magnetic aerosols, loaded with therapeutic substances, at target sites within the respiratory system. In this workflow, the magnet plays a major role.

The experiments conducted by the research partners in the scope of this project involved a strong electromagnet that has been conceived and developed by the company Siemens Medical Solutions (Erlangen, Germany) (C. Alexiou, 2006). To accompany these experiments on a simulative basis, a 3D modeling of the used magnet has been realized using the construction and simulation software COMSOL Multiphysics (Version 3.5a). A particularly relevant application of this numerical reconstruction was the extraction of magnetic field and field gradient data that were used in the computation of the aerosol trajectories and deposition in Fluent.

The Siemens magnet was conceived in a way to guarantee a magnetic field gradient with values above 10 T/m in a zone surrounding the magnet tip covering a volume of 2x2x2 cm^3. Figure 28 shows the geometry and mesh of the simulated electromagnet.

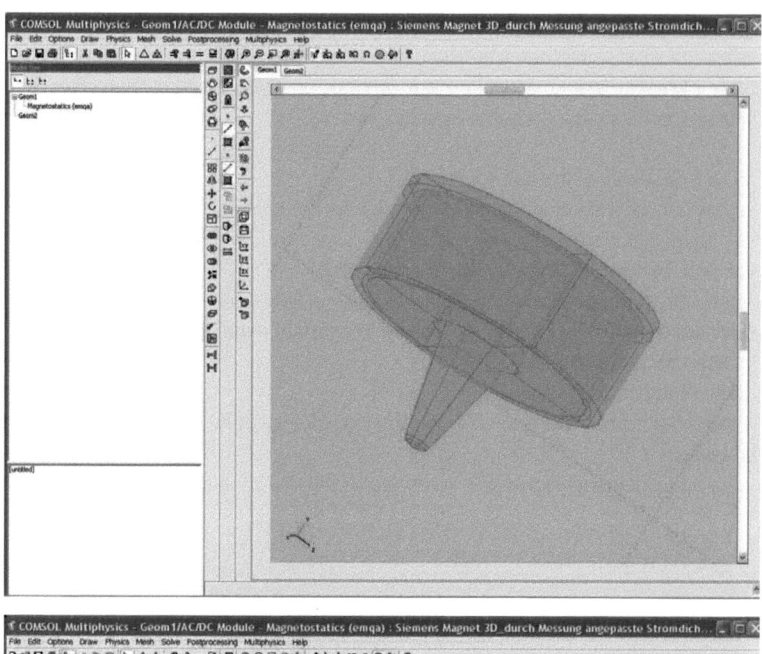
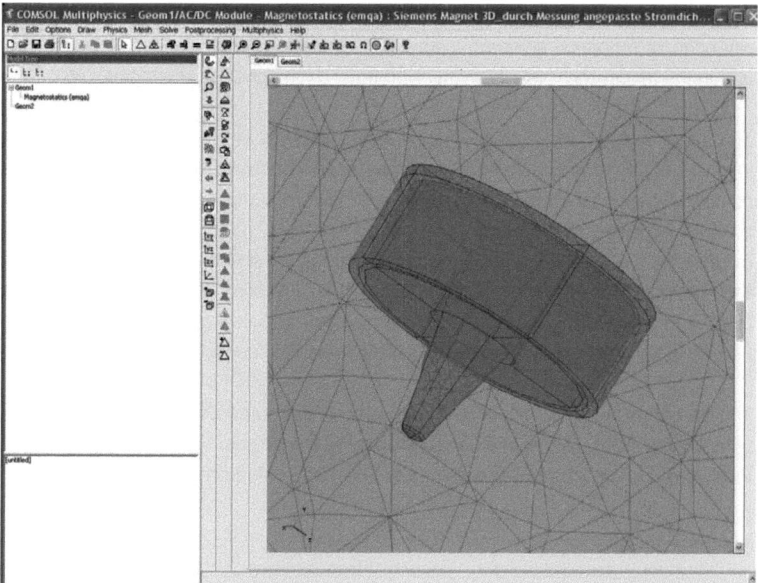

Figure 28: Geometry (top) and Mesh (bottom) of the simulated Siemens magnet

Using the post-processing options of COMSOL Multiphysics, the magnetic flux density and the magnetic field gradient were plotted in the relevant space under the magnetic tip, as illustrated by Figures 29 and 30.

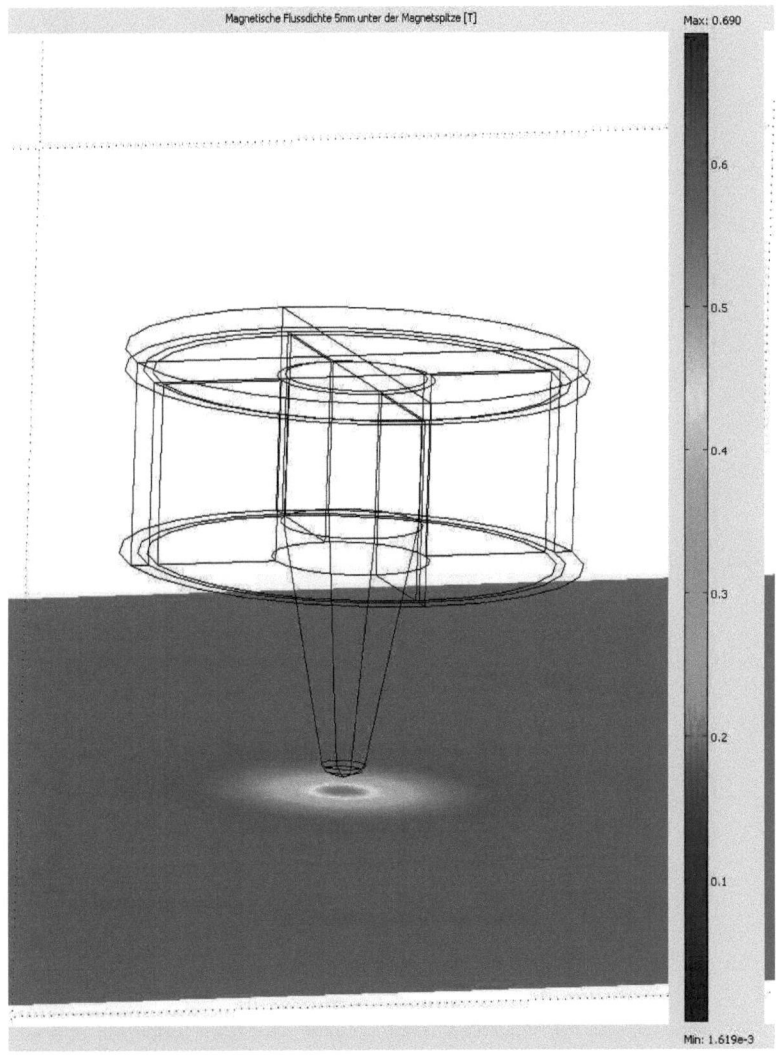

Figure 29: Magnetic flux density in a layer 2 mm under magnet tip.

Figure 30: Magnetic field gradient (bottom) in a layer 2 mm under magnet tip.

To further validate the simulation, measurements have been conducted to assess the magnetic flux density as well as the magnetic field gradient underneath the magnet tip. The results are reported in Figure 31.

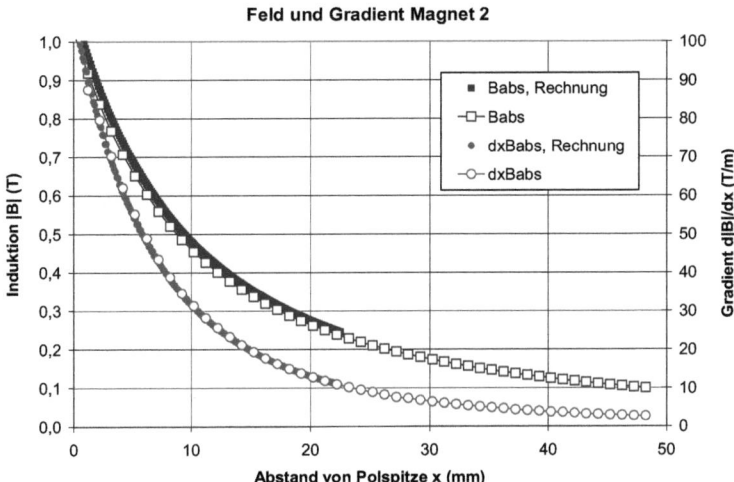

Figure 31: Measured magnetic field data (flux density and field gradient) – Siemens magnet (C. Alexiou, 2006).

As the simulation of the Siemens magnet was performed in the three dimensional coordinate system, the results of the simulated magnetic field have been compared and validated with the measured values in a vertical plane crossing the magnet tip, so that a transition between experimental and computed results was assured. The results from the simulation are displayed in Figure 32.

To assign a magnetic force to the aerosol particles, the computed values of the magnetic field gradient were extracted from COMSOL Multiphysics and transferred to Fluent via a so called User Defined Function (UDF). Hereby, the three dimensional gradient components were saved as a variable from which the corresponding values can be retrieved at each step of the simulation. The magnetic force is then calculated according to the formula (1) and used to determine the acceleration of a given aerosol particle following Newton's second law of motion.

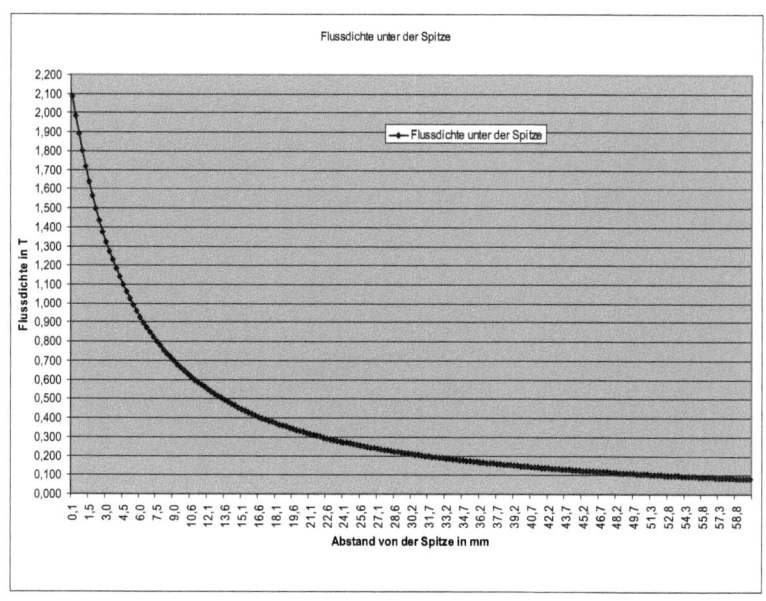

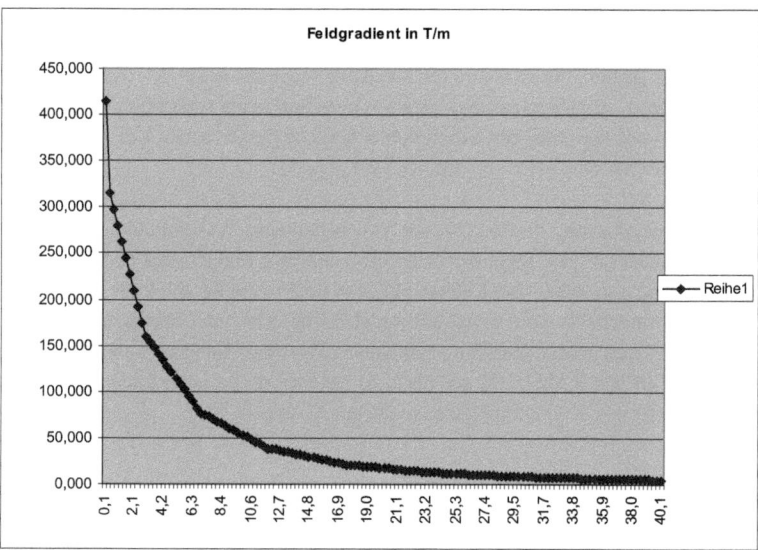

Figure 32: Simulated magnetic flux density (top) and simulated magnetic field gradient (bottom) for distances given in mm under magnet tip.

The magnetic field data extracted from COMSOL are encapsulated in a six column matrix with the first three columns dedicated to the coordinates in space of the assessed points and the last three filled with the gradient values. Through the repartition of the relevant space occupied by the lung model into equidistant layers, a mesh containing a defined number of space points was created, with gradient field and therefore magnetic force values computed in each single point. A high mesh resolution of 200x200x200 was adopted to guarantee a high accuracy level in determining the involved magnetic forces.

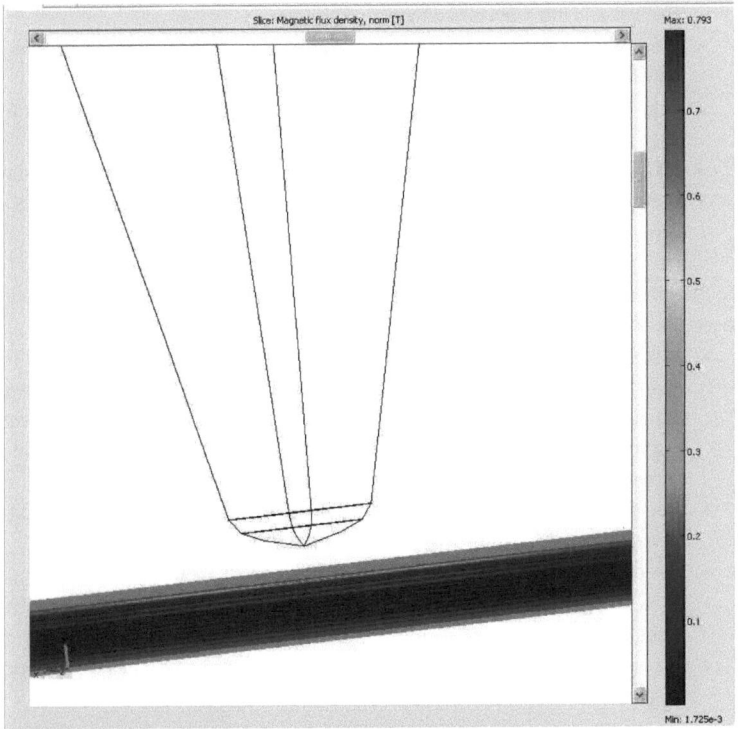

Figure 33: Illustration of extraction layers where the magnetic field data are plotted under the magnet tip.

The simulation of the aerosol trajectories was performed using 20,000 aerosols, injected following a defined inlet particle distribution calculated for a fully established inflow profile (Issa, 2010). An iterative computation was performed for each aerosol based on the balance of forces acting on it. The magnet tip was positioned at the level of the branch "lr" (Figure 24) in the left lung to avoid the intrinsic influence of the higher air flow rate reaching the right lung due to the anatomical asymmetry.

The most important results that were achieved can be described in relation to the intubation effect and the magnetic field influence.

On the one hand, simulative evidence was established in case of intubation that (Issa, 2010):

- the assumed laminar air flow features turbulences in the region of the trachea and the main bifurcation.
- there was no significant difference in the aerosol deposition rate when using different tube ending forms
- over 50 % of aerosols are trapped in the main bifurcation opposite to the inlet
- for aerosols with a 10 μm diameter, the deposition rate increases significantly
- the magnetic field increases the deposition rate for aerosol diameters up to 5 μm

On the other hand, the application of a magnetic field gradient led to the following observations (Issa, 2010):

- a positive effect could be registered when varying the activation time of the magnetic field. In the inspiration phase, this triggering generated the most significant effect when it was synchronized with the aerosol front approaching the magnet tip area. For the expiration phase, an optimal activation should be considered from $t = 0$ s.
- the deposition rate decreases with higher air flow rates, as the airborne aerosols rapidly pass through the magnetic trapping zone.
- the deposition rate decreases for greater aerosol diameters
- for a successful therapy, higher air flow rates have to be avoided and aerosol sizes limited.
- varying the incidence angle of the magnet tip results in a changing deposition rate.

Figure 34 shows the concentration of MNP loaded aerosols subjected to a magnetic field gradient positioned over the left lung, with the magnet tip placed orthogonally to the plane of the figure right above the left bifurcation (Bif_L).

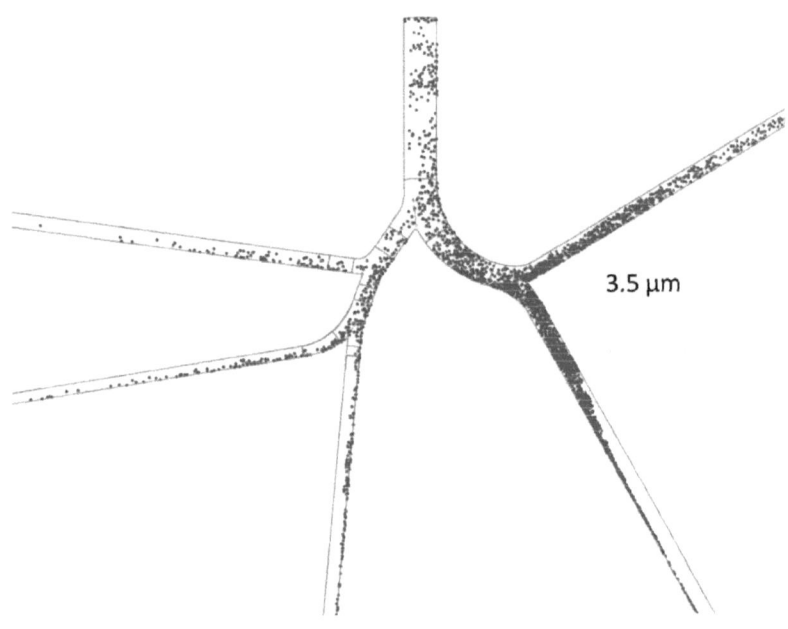

Figure 34: Retention effect on MNP loaded aerosols subjected to a magnetic field gradient positioned over the left side of the lung (Issa, 2010).

For selected parameters, the diagrams in Figure 35 reveal the influence of the variation of the observed quantity on the corresponding deposition rate (in percentage of the total number of inhaled aerosols). The main focus of the simulative study was accorded to the magnetic field activation time, the magnetic tip angle with respect to the plane of the target segment, as well as the aerosol size (diameter) and air flow rate.

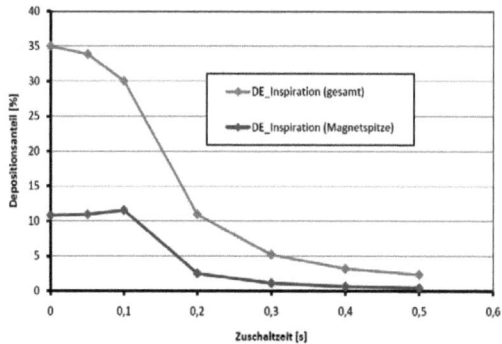

A: Effect of magnetic field activation time on aerosol deposition rate during inspiration (green line for total deposition, red line for deposition in the airway segment under magnetic tip)

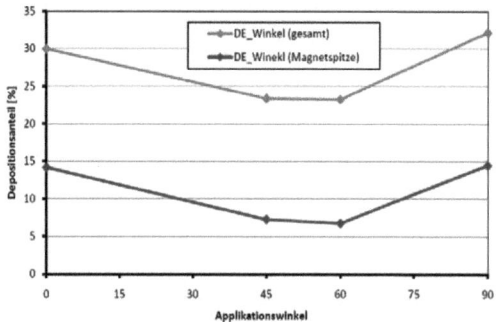

B: Effect of magnetic tip angle above the target segment of the lungs on aerosol deposition rate (green line for total deposition, red line for deposition in the airway segment under magnetic tip)

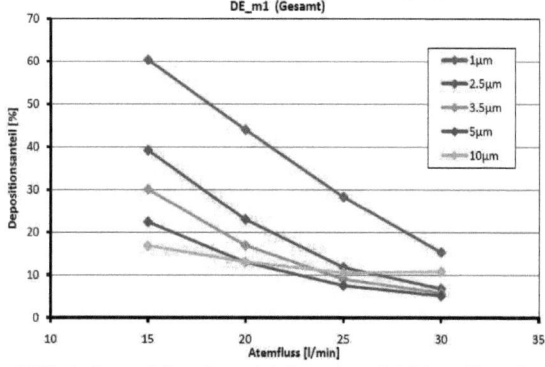

C: Effect of aerosol diameter and air flow rate on total deposition rate

Figure 35: Selected results of particle behavior and sedimentation in the proposed lung model under magnetic field activation (Issa, 2010)

In conclusion, it can be confirmed that an optimal constellation of parameters exists to best plan a Lung Drug Targeting procedure, based on simulative studies. For instance, the main requisites for a maximum aerosol deposition rate include an air flow rate of 15 l/min, a magnet tip placed at an angle of 90° relatively to the vertical reference, and a triggering of the magnetic field 0.1 s after aerosol injection into tracheal inlet. An immediate consequence of this multiparametric study is the possibility to optimize pre-clinical experimentations with a reduced number of involved animals.

4.2.3 The effects of intubation

The multi-parametric study demonstrated that the intubation in the scope of Lung Drug Targeting has a tremendous impact on the aerosol deposition rate. This can be clearly observed in Figure 36, where a 20-fold higher trapping efficiency is achieved when using a tube for inhalation. This effect is furthermore mainly due to the positioning of the tube in the trachea, and less to the shape of the tube extremity. For instance, a flat or sharp (pointy) tube profile would both lead to the same deposition rates.

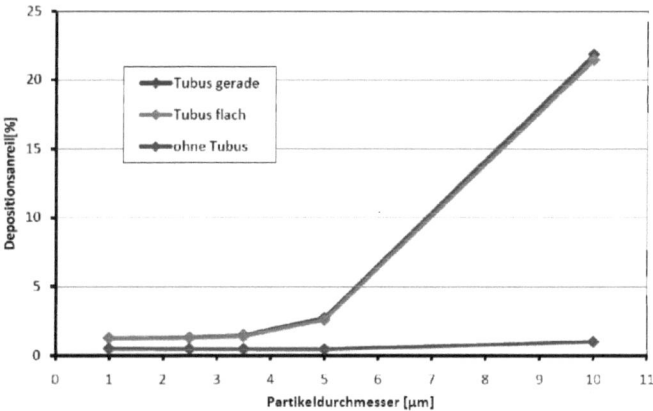

Figure 36: Effect of intubation on aerosol deposition rate, with varying aerosol diameter (blue line: no intubation, green and red lines using tubes of different extremity shapes) (Issa, 2010)

Beyond the examined cases, deep intubation can be considered, which would lead to a more efficient, distal delivery of the aerosols into the lungs. Intuitively, this option is completely opposed to the envisaged clinical scenario for LDT application which is based on the natural breathing of the patient. On the other hand, a natural breathing would inevitably lead to the unwanted deposition of the inhaled aerosols in the lower mouth and upper pharynx region.

4.2.4 Pre-clinical evaluation of Lung Drug Targeting

In the course of the research program to investigate Magnetic Drug Targeting and Lung Drug Targeting conducted by the group of partners "Nanomagnetomedizin" and granted by the German Ministry of Education and Research, a special attention has been dedicated to pre-clinical experimentation in the big animal model. In the scope of this thesis, support has been provided to the team around Dr. Carsten Rudolph from the Research Center of the Pediatric Clinic and Polyclinic of the „Dr. von Haunersches Kinderspital" in order to optimize the experimentation of MDT on animal lungs. These studies involved the intubation and ventilation of experimentation pigs inhaling magnetic aerosols that are retained at a given location in the lungs by a strong magnetic field. The goal is to increase the concentration of the deposited active agents in the corresponding area (Nanomagnetomedizin, 2009).

The aerosols were loaded with the magnetic nanoparticles fluidMAG-D of the company Chemicell GmbH (Berlin, Germany) as well as radioactive iodine I^{125} for marking, and directed to the upper respiratory pathways of the animals. The pigs were kept under general anesthesia during the treatment. The previously described Siemens magnet was used and placed above the targeted thorax compartment (Nanomagnetomedizin, 2009).

Figure 37 shows the experimental setup used in these studies and involving the intubation device with the nebulizer, as well as the electromagnet.

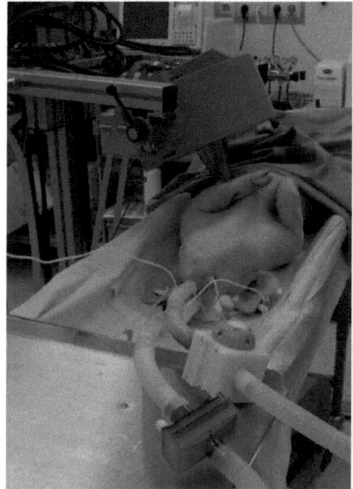 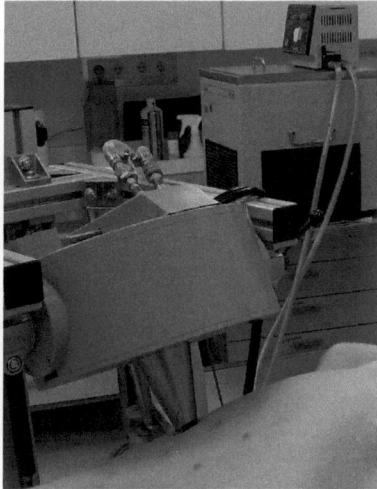

Figure 37: Experimental setup for the preclinical Lung Drug Targeting studies conducted on the big animal model (pig) and featuring the positioning of the electromagnet tip.

During experimentation, a cooling system was used to keep the temperature of the electromagnet at 25°C. After the application of 10 respiration cycles, the thorax of the animal was imaged with a gamma camera then with a CT scanner. Both acquisitions were then combined to obtain a full functional cartography of the animal. In part of the acquisitions, a PET-CT has been used.

Figure 38 exemplarily depicts the distribution of the radioactively marked nanoparticles in the lungs of the treated animal for one single treatment. The bright spots in the left lung reveal a significant concentration of aerosols in the corresponding regions, particularly at the level of the magnet tip, whereas the rest of the lungs is clearly unaltered.

For the described case, the particle concentration and distribution has been assessed quantitatively. By way of example, the case documented in Figure 38 features a repartition of the detected radioactive particles as follows:
- Left Lung: 589196 counts, i.e. 63.9% of total number of aerosols deposited
- Right Lung: 332847 counts, corresponding to 36.1% of deposited aerosols.

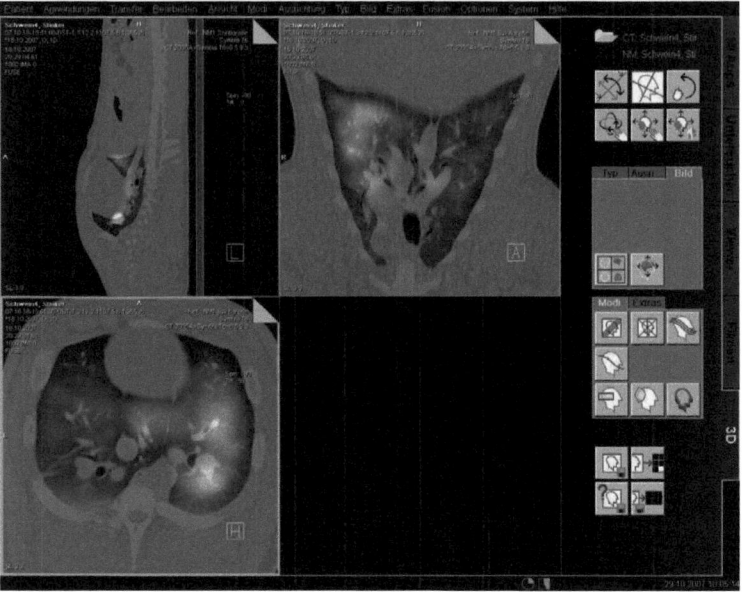

Figure 38: Distribution of radioactively marked nanoparticles in the lungs of a pig after LDT application

The obtained results were in accordance with the outlooks predicted by literature, analytical analysis, and simulation. More generally, the animal experiments led to the optimization of the ventilation parameters and the properties of the applied nanoparticles, which resulted in the successful demonstration of the retention effect in the targeted lung region. SPECT-CT images as well as histological investigations showed a significant accumulation of magnetic nanoparticles at target site.

In parallel to the evidences achieved through these experimentations on the big animal model, it is of great utility to still consider the aspects that differ from human application. In fact, differences and variations exist already within the lung structures of a given species, let alone when transiting from one species to another. Therefore a direct comparison between experimental results in pig lungs and human lungs should be avoided, especially in terms of flow conditions and particle deposition (Nikivorov A.I., 1985). Moreover, the swine presents a number of particularities in its anatomical structures that clearly differ from human physiology. For example, the trachea of a pig is relatively longer and more curved when compared to a human trachea. In addition, the bronchial tree of a swine presents a first minor branch distally from the main bifurcation, which has an impact on the air flow distribution and velocities, as well as the therein suspended particle behavior that can not be neglected. These anatomical particularities in the pig can easily lead to a higher deposition rate of the applied aerosols at the level of the upper pulmonary pathways due to eddies and turbulences.

A reconstruction of the upper respiratory tract of three different pigs used in the experiments is shown in Figure 39 and clearly reveals this anatomical signature.

Figure 39: Reconstruction of the upper tracheo-bronchial tree of three different pigs treated in the pre-clinical evaluation and exhibiting a first minor bifurcation before the main bronchi

Beyond these pertinent observations, the modeling of the different geometries of the involved anatomies for simulative studies also allowed a deeper understanding of the ventilation conditions. As a matter of fact, the positioning of the tube in the upper region of the trachea could be visualized (Figure 40) under real treatment conditions and revealed a tendency to direct the inhaled airflow towards the inner walls of the windpipe, causing -in a number of treated animals- an unusually increased aerosol deposition in that area.

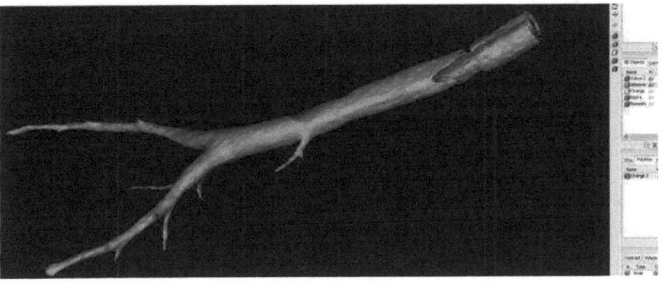

Figure 40: Reconstruction of the upper bronchial tree of an experimentation pig including the tube used for ventilation. The inclination of the tube explains the unusually increased aerosol deposition in the upper trachea.

4.3 Breath-synchronous lung drug targeting

Lung Drug Targeting is certainly one of the most promising methods to efficiently treat diseases of the respiratory system, such as carcinomas in the lungs. However, first implementations of this technique, as described in the pre-clinical studies presented in chapter 4.2.4 still present room for improvement. One of the suboptimal aspects of the current treatment scenario is the facto f assuming a continuous application of the magnetic field all through the inspiration and expiration phases of the artificial respiratory act that supplies the patient.

During experimental investigations, it was observed that applying the field this way forced the magnetic aerosols to sediment at regions far from the target site, mainly in the trachea and main bronchi, because of the force inducing magnetic field gradients that are present over the whole field application area.

Considering these effects led to the development of an optimized approach consisting in:
- punctually generating the aerosol cloud exactly at the beginning of the inspiration phase, which would propel the particles to the deepest parts of the lungs and therefore to the targeted cells as well
- synchronizing the magnetic field activation with the breathing process, so as to trigger the magnetic trapping right when the aerosols have reached the target site

Further elaboration on this approach involves analyzing the relevant respiration parameters such as pressure and flow in order to detect the end of the inspiration phase and trigger the magnet exactly at the point in time, when particles have reached the deepest alveoli at target site, and are no more subjected to forces due to the streaming. The magnetic field is then held on during the expiration phase to assure the retention of the aerosols at the targeted sites, which increases the efficiency and focality of the treatment. This way, only target cells are subjected to the deposition of the drug carrying aerosols, while the other healthy regions of the lungs remain unaltered by side effects (Dahmani Ch., 2008).

This inevitable improvement of the concept of Lung Drug Targeting through the synchronization with the breathing process is summarized in the diagram shown in Figure 41.

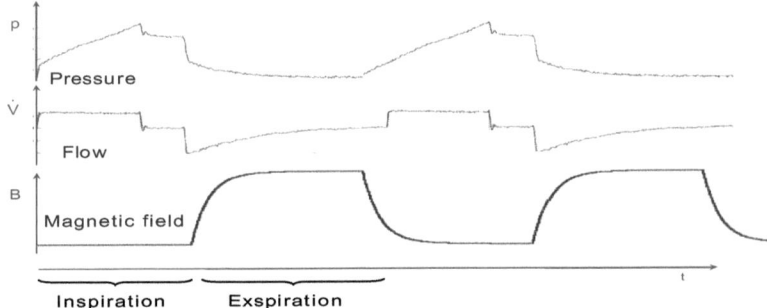

Figure 41: Time schedule for magnetic field activation in an optimized, breath-synchronous Lung Drug Targeting procedure (Dahmani Ch., 2008)

5 Cellular uptake of nanoparticles following successful targeting

The ultimate goal of Magnetic Drug Targeting is to selectively bring a substance capable of deploying therapeutic activity to an ailment site in the body, using natural circulation pathways, such as blood vessels or respiratory pathways. Yet, successful therapy is only achieved, when the therapeutic load carried by the guided nanoparticles is internalized into the target cells and triggers a healing process, either through necrosis -when the intended effect is eliminating the cell-, or through the initiation of a therapeutic protein production activity within cells, in the case of gene therapy.
In both variations, cellular uptake of the internalized bioactive substances and/or genetic material is indispensable.

Several techniques have been developed to overcome cell membranes and enhance substance delivery into cells. This process is called cell transfection, or transduction, when it involves the use of viruses or viral shells as mediators or carriers for the transported load. Usually, these two approaches are referred to as viral and non-viral gene delivery. In fact, viruses are very efficient in delivering genes into cells, but they lack safety. This motivated researchers to explore alternative ways of overcoming cellular barriers that –even if not as efficient as viruses- hold the promise of safety and reproducibility. Among other advances, self-assembled nanosystems have been investigated for targeting subcellular organelles (Torchilin, 2006).

Applying controlled and enhanced transfection methods presents several advantages, including faster operation and higher rates, and has become routine in daily biological work, making it for instance possible or considerably easier to produce proteins for clinical or research applications, to add genetic markers to cell lines and more generally to study DNA replication, recombination and mutation.

In this section of the study, ways of applying improved transfection techniques to assure the cellular uptake of the delivered bioactive substance at the magnetically targeted site following successful MDT are explored.

5.1 Enhancing cell permeability with static magnetic fields: Magnetofection™

Transfection, defined as the non-viral process of introducing substances into cells, plays a significant role in molecular biology and genetic engineering. It enables the mass manufacturing of pharmacological or biological beneficial products, and leads to an industrial exploitation of genetically modified cells or bacteria (C. Bergemann, 1999).

Transfection is also indispensable in magnetic drug targeting and gene therapy where nucleic acids (genetic material) constitute the transported load, comprising plasmid DNA or siRNA fragments, or proteins, e.g. antibodies, and have to be selectively conducted to the subcellular compartments.

When the cargo comprising the genetic material is delivered into the cell, the proteins encoded therein are expressed, i.e. replicated by the cell and amplified, deploying thus a therapeutic effect.

Especially in drug delivery, the challenge goes beyond ensuring the uptake of these substances into the cells, as the real target sites are mostly in the intracellular space (cytoplasm) or even in the nucleus.

To make molecules traverse the cell membrane, many techniques have been explored and applied. For instance, electroporation, involving exposure to high voltage electric fields, sonoporation, based on the application of ultrasound waves, or particle bombardment have been reported to mediate cell transfection (Chao-Bin Chen, 2009).

The generality that can be observed from these different approaches is the necessity to create transient openings or pores in the cell membrane, allowing for material transport and uptake. This process should be in analogy with the natural cellular uptake mechanisms for nanoparticles and macromolecules, involving pinocytosis, endocytosis, and receptor-mediated endocytosis (Challa S.S.R. Kumar, 2005). Figure 42 illustrates these different techniques.

Living cells possess plasma membranes that are characterized by a dynamic, lipophilic structure. It follows from this that the entry of large, hydrophilic, or charged molecules into intracellular space is limited. Because most genetic molecules are not only charged but also relatively large in size, they cannot cross the cell membrane on their own, but have to be delivered to the cytoplasm using an accordingly sophisticated transport system.

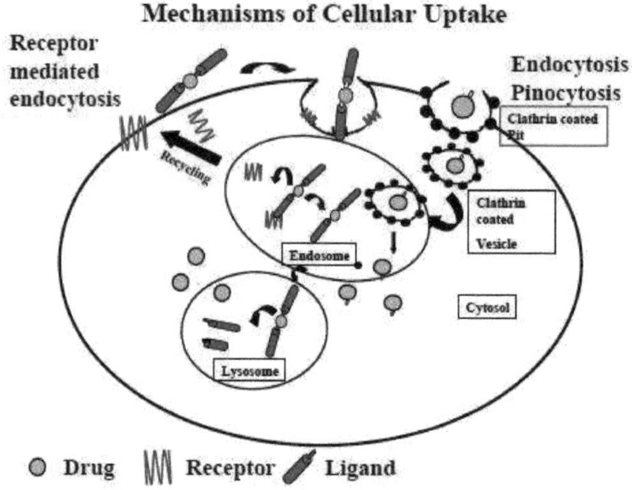

Figure 42: Natural mechanisms of cellular drug uptake into tumor tissue, as illustrated in (Challa S.S.R. Kumar, 2005)

A particularly successful development responding to this challenge has been the assembling of synthetic or non-viral gene delivery systems. Especially the latter present pertinent advantages, such as their ease of use and simplicity of production in important quantities. Non-viral methods also prevent the triggering of nonspecific inflammations followed by unintended immune responses (Ikramy A. Khalil, 2006).

In addition to the increasing tendency of using non-viral gene delivery systems to perform the biotechnological maneuver of transfection, the use of nanoparticles combined with magnetic fields has also gained in importance over the last years.

In fact, researchers exploit magnetized nanoparticles not only for the guidance and control of substances within blood vessels and respiratory pathways, or for cellular and macromolecular separation processes (Wei, Lee, & Lai, 2009) (José Luis Corchero, 2007), but also to reproducibly enhance the rate of cell transfection.
It is here of great utility to underline, that the process of cellular uptake is mainly limited by the insufficient contact between traditional delivery systems and target cells (Tristan Montier, 2008), leading to sub-optimal transfection rates.

By adequately coating and loading magnetically activated nanoparticles with DNA plasmids and transporting the resulting complexes through the cellular membrane to the inner parts of cells (C. Bergemann, 1999), Plank et al. were among the first researchers to report successful implementations of this magnetically mediated transfection and established this method in 2000 as the technique of Magnetofection™(Plank C., 2000).

Figure 43 shows the principle of standard magnetofection.

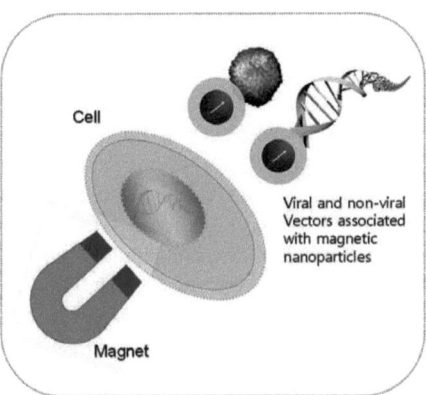

Figure 43: Illustration of the Magnetofection technique, consisting in the nucleic acid delivery to target cells guided and enhanced by the action of magnetic gradient fields on gene vectors associated with magnetic particles.

The main strength of this technique is the enhanced contact of transported vectors with target cells, as the used magnetic nanoparticles associated with nucleic acids are prevented from diffusion and forced to sediment onto the surface of cells within short times (Vainauska D, 2012).
In accordance with this explanation, Magnetofection has shown tremendous improvements in transfection efficiency of reporter genes, as compared to conventional transfection methods (Plank C, 2003). These improvements have been reported in vitro and in vivo (F. Scherer, 2002), and for different gene types and cell lines, including neuronal and glial cells, tumor and embryonic cells, as well as endothelial and epithelial cells (José I. Schwerdt, 2012).

One further major advantage of the magnetically actuated transfection is the intrinsic ability to use the magnetic field to separate the transfected cells -containing for instance magnetic nanoparticles- from the non-transfected cells, avoiding, so, cumbersome and costly alternative measures (C. Bergemann, 1999).

5.2 Cell transfection with dynamic magnetic fields

The achievements realized through the investigation of magnetofection represent an enormous potential for this technique to complete the technology of Magnetic Drug Targeting. The combination of both methods is intuitive, as in each of them, magnetic fields are used to exert attraction forces on nanoparticles carrying therapeutic loads.

Provided the external magnetic fields are properly activated and concentrated to retain the nanocarriers, they can also be accordingly focused to enhance cellular uptake of the transported bioactive agents. Consistent with these outlooks, it can be imagined that the therapeutic load carried by the nanoparticles comprises genes coding for a certain therapeutic molecule. Once this genetic material is delivered to the intracellular milieu, the target cells start producing their "medicine" by themselves. Such a promising scenario highly depends on the efficiency of the transfection method used. Therefore, it is of great interest to investigate ways of improving cellular uptake mechanisms for magnetically targeted substances.

In this chapter, approaches to enhance the efficiency of magnet assisted transfection are explored. These are based on the extension of the classical static magnetic fields -proven to accelerate vector deposition at the cell surface- to dynamic magnetic gradient fields.

In fact, time-varied magnetic fields have been reported in an exploratory phase to facilitate the transport of magnetic nanoparticles in artificial cytoplasm-like media, such as viscous gels (Hanjie Wang, 2010), and recently, alternating or pulsating fields were reported to definitely increase magnetofection efficiency (Sarah W. Kamau, 2006). For instance, Kamau et al. used a Dynamic-Marker magnetic device to generate a sinusoidal wave exhibiting a peak amplitude of 27 mT and a field gradient of 10 T/m perpendicular to a cell culture exposed on a well plate. The frequency of this component of the alternating field was 50 Hz. Additionally the "magnetic waves" had parallel components to the plate's surface that oscillated at 0.75 Hz. Kamau et. al reached an average increase of up to a 2-fold in the transfection rates compared with an exposure to a static magnetic field, and further reported improved transfection in the case of primary cells, including synoviocytes, chondrocytes, osteoblasts, melanocytes, macrophages, lung and embryonic fibroblasts (Sarah W. Kamau Chapman, 2008). Based on the same principle but using a different setup, Chen et al. exposed mammalian cells and Escherichia coli (adherent and in suspension) to pulsed magnetic fields of millisecond duration and featuring amplitudes of 0.6 T, and achieved in defined cases a more rapid transfection. Given the strong dependency of the transfection efficiency on the used magnetic particles, the number of magnetic field pulses and the DNA dose, they reported the most significant increase to be reached through pulsing three times with magnetic fields of up to 2.15 T (Chao-Bin Chen, 2009) (Ji-Yao Chen, 2006).

A particularly convenient implementation of the same concept was realized by Jon Dobson's team who developed and commercialized a system generating alternating fields based on a conventional magnetic plate array moved below a cell culture support at a low frequency and small amplitude. Improvements in the transfection efficiency of cell lines and primary cells could be achieved using this setup (Pickard M, 2010) (Fouriki A, 2012) (McBain SC, 2008).

In the frame of this thesis, a magnetic system following the same principle of pulsating magnetic fields, yet exploiting a different concept, has been developed.

5.2.1 Rotational magnetic system and pulsating fields for cell transfection

As reported in the literature, most of the systems used to generate magnetic pulses for biological applications are based on electromagnets. Yet this concept presents major drawbacks with regards to highly sensitive cell experiments. For instance, Eddy currents and the resulting heating in electromagnets generate a variation in the electric current flowing through their windings (Eberhard Kallenbach, 2003) (C. Alexiou, 2006), which not only leads to instabilities in the developed magnetic field and the overall experimental parameters but also represents a danger and a disturbing factor for the treated cells.

To exclude the drawbacks of electromagnets in the process of enhancing cell transfection with magnetic pulses, a new concept needed to be elaborated, that would only involve permanent magnets and still generate the necessary variable field. To be able to deliver pulsing fields with these magnets, a comparable mechanical system that assures a very rapid transient change in the magnetic flux density surrounding the nanoparticles has been conceived.

An eventually translational movement of a permanent magnet would not be as fast as needed and therefore could not deliver the needed changes in terms of field strength. Therefore, the approach capable of generating the intended pulses involved a rotating system which keeps the flux density in the targeted region varying rapidly over the time. The estimated pulse magnitude should hereby exceed $6*10-3$ to $7*10-2$ T and the pulse duration should ideally be in the range of 10^{-5}s to 10^{-1}s (Timothy E. Vaughan, 1996).

The concept was implemented in simulations to predict the behaviour of the flux density in the targeted sites and adjust frequencies and amplitudes.

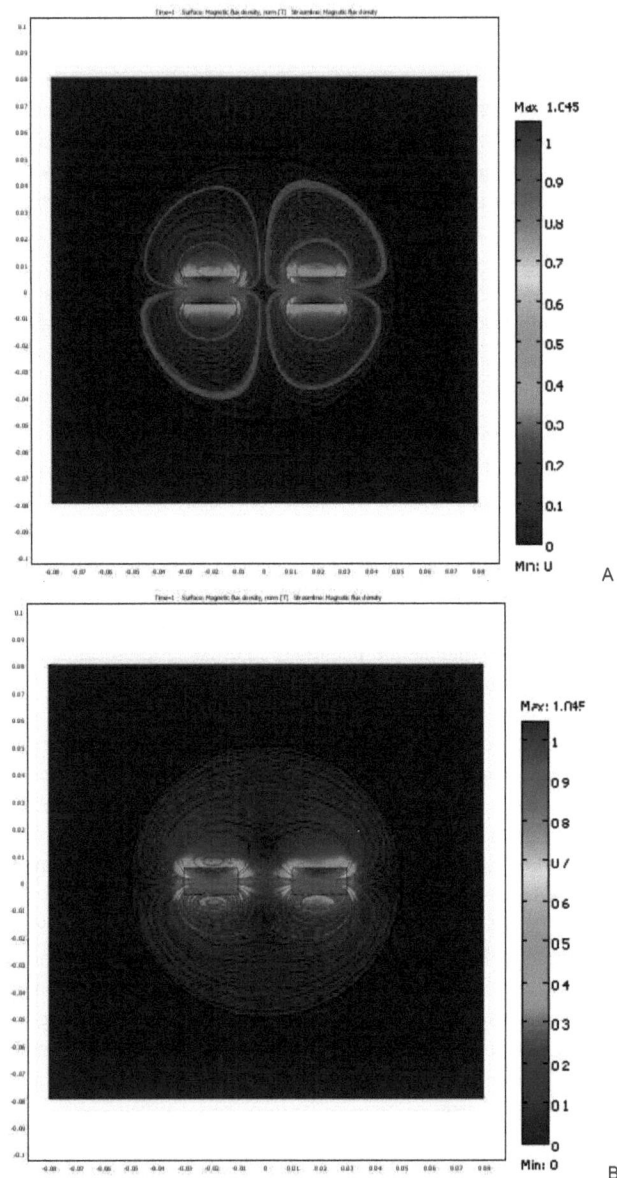

Figure 44: Magnetic flux density distribution (in T) and magnetic field lines of the simulated magnet dispositions for a magnetization along width of rod, parallel (A), anti-parallel (B)

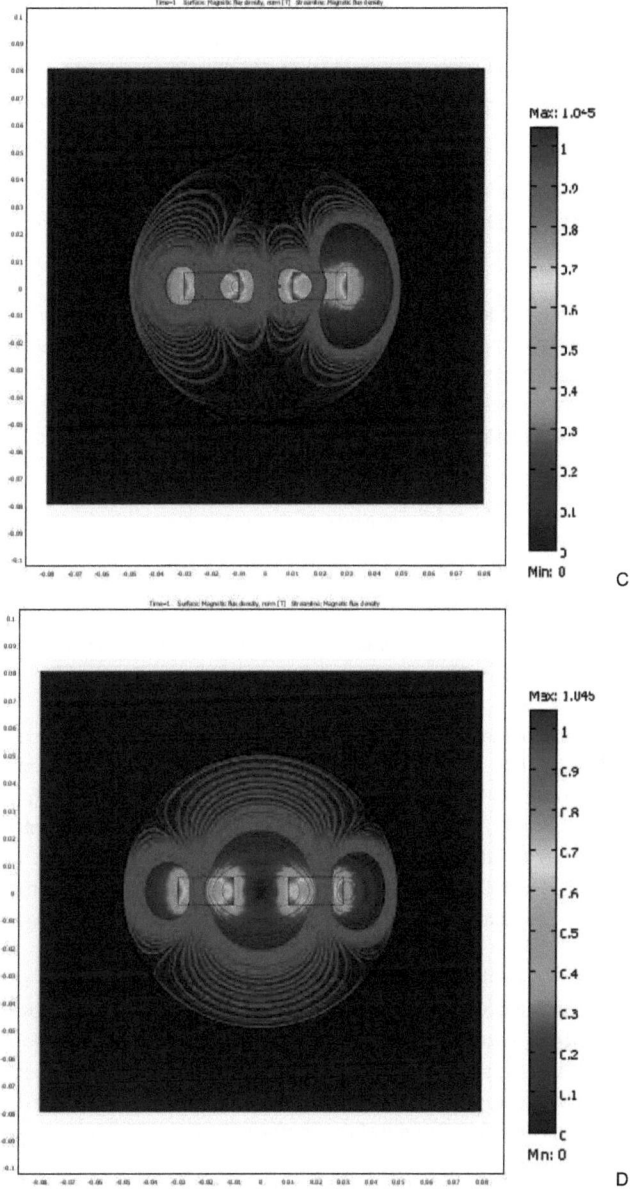

Figure 45: Magnetic flux density distribution (in T) and magnetic field lines of the simulated magnet dispositions for a magnetization along depth of rod, parallel (C), anti-parallel (D)

The conceived system was therefore modelled in COMSOL Multiphysics and consisted of two permanent magnet rods (10x20x120 mm) rotating counter-clockwise and separated by a distance of 20 mm. Each of them had a remanent flux density of 1.43T.

The two magnet rods were placed following 4 possible arrangements, as depicted in Figures 44 and 45, with a magnetization along the width and along the depth of the rod, in parallel and anti-parallel direction.

The rotation was modeled using a deformed mesh application mode, in which the center part of the geometry, containing the rotating magnets and the space between and surrounding them (indicated through the index "rot"), rotates with a rotation transformation relative to the default coordinate system (indicated through the index "stat"). The rotation of the deformed mesh is defined by the transformation

$$\begin{pmatrix} x_{rot} \\ y_{rot} \end{pmatrix} = \begin{bmatrix} \cos(wt) & -\sin(wt) \\ \sin(wt) & \cos(wt) \end{bmatrix} \cdot \begin{pmatrix} x_{stat} \\ y_{stat} \end{pmatrix}$$

where ω is the angular frequency.

The rotating magnetization was also implemented following the same transformation (Comsol Multiphysics, 2009).

The results of the simulation were then adjusted according to the constructed system involving two permanent magnet cuboids made out of Neodymium-Iron-Boron (NdFeB) with a 10 micron coating of Nickel (Ni+Cu+Ni) and having a residual induction of B_r = 1.43T. The magnet was controlled with a brushless motor system that rotated up to 4000 turns per minute, which corresponds to a maximum frequency of 2 x 4000/60 s = 133.33 Hz; a sample treated with this magnet constellation would experience two pulses with every revolution of the magnet pair.

As shown in Figure 46, a well holder was constructed to suit the used "Nunc BreakApart" detachable wells and was positioned along the axial axis of the magnetic system. Two variant well holder platforms were used, allowing for two placement possibilities for the wells, with distances r from the system centre equal to 4 and 7.5 mm, corresponding to magnetic flux densities of B_1 = 190 mT and B_2 = 310 mT, respectively, as measured at the central points of the wells.

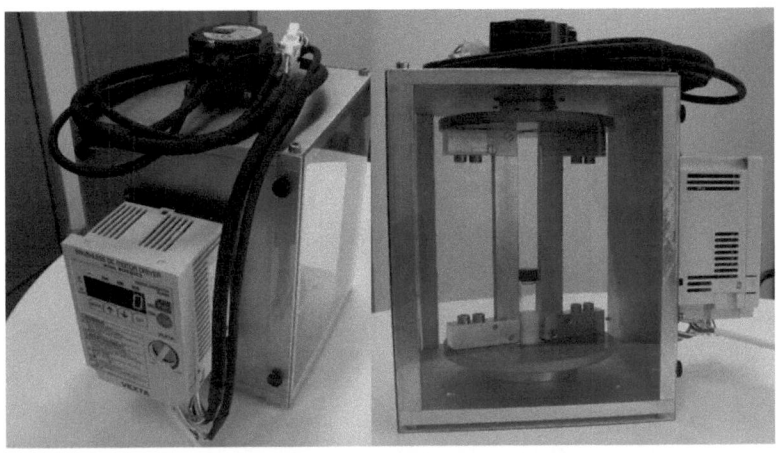
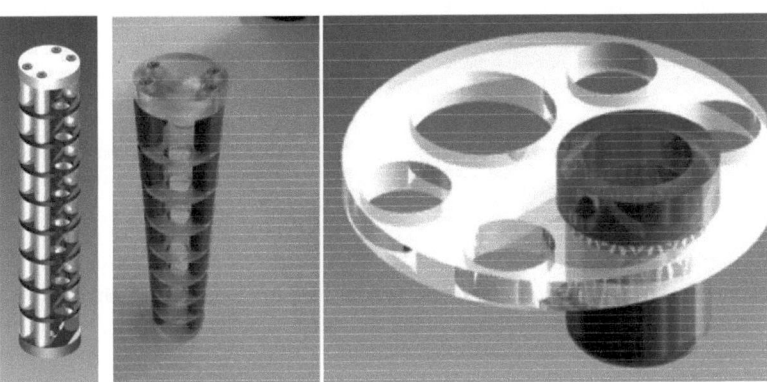

Figure 46: Top: the constructed rotating magnetic system with brushless motor, bottom: well holding disc (right) and multilevel well holder (left) for magnetic exposure of cells

The developed rotational magnetic system produced the expected sinusoidal pulse forms at the chosen reference points. The diagrams in Figure 47 show the simulated magnetic flux density [in T] and the field gradient [in T/m] at a distance of 7.5 mm from the system centerline, featuring an amplitude of 310 mT and a frequency of 10 Hz. As the direct measurement of these values over the time in the space between the cuboids was not possible during system rotation, the simulation was indispensable to determine the magnetic field behavior at the reference points.

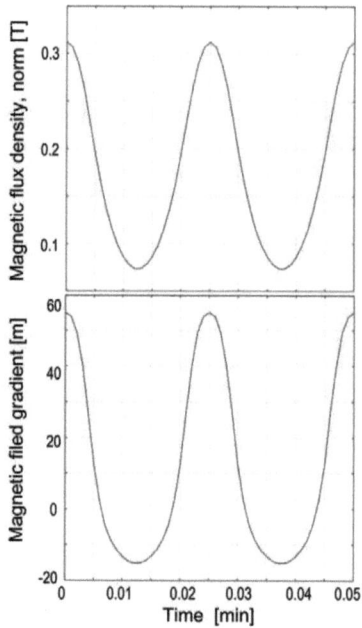

Figure 47: Magnetic flux density and magnetic field gradient at a distance r = 7.5 mm from the system center and a frequency f = 10 Hz

5.2.2 Experiments

With the developed magnetic system, cell transfection experiments were conducted on adherent as well as suspension cells, using magnetic nanoparticle/plasmid complexes.

The used magnetic nanoparticles:
For the transfection experiments, two kinds of nanoparticles have been used. For the trials involving adherent cells, the PEI-Mag2 magnetic nanoparticles, as described in Chapter 2.1 were used, whereas for the suspension cells, SOMag5 core/shell-type magnetic nanoparticles were applied. These had an iron oxide core with an average crystallite size of 6.8 nm and a silica oxide coating with surface phosphonate groups; the surface phosphonate groups were formed by the co-condensation of tetraethyl orthosilicate and 3-(trihydroxysilyl)propyl methylphosphonate. The mean hydrodynamic diameter and zeta potential of the MNPs suspended in water were measured to be 40±14 nm and 38.0±2.0 mV, respectively. The saturation

magnetisation per unit of iron weight at 298 K was 94 emu/g iron. The aqueous MNP suspensions were sterilised using ^{60}Co gamma-irradiation (25 kGy).

The cell culture:
For the experiments in adherent cells, NCI-H441 human pulmonary epithelial cells were cultured. Referred to as H441 cells, they were derived from a papillary carcinoma of the lungs (ATCC, cat. no. HTB-174). The used culture medium comprised modified RPMI 1640 medium with 2 mM L-glutamine, 10 mM HEPES, 1 mM sodium pyruvate, 4.5 gl^{-1} glucose, and 1.5 gl^{-1} sodium bicarbonate supplemented with 10% heat-inactivated FCS, 100 U/ml penicillin, and 100 µg/ml streptomycin. The cells were grown at 37°C in a humidified atmosphere containing 5% CO_2.
For the experiments with suspended cells, Jurkat T cells were obtained from DSMZ (Deutsche Sammlung von Mikroorganismen und Zellkulturen GmbH, cat no. ACC 282) and maintained at 37°C and 5% CO_2 in RPMI 1640 medium (Gibco-BRL, Eggenstein, Germany) supplemented with 10% foetal calf serum (FCS), 2 mM D-glutamine, 100 U/ml penicillin and 100 µg/ml streptomycin (complete medium) at a density of 0.75 to 1.5 million cells/ml. Every 2–3 days, at a density of approximately 1.5 million cells/ml, the cells were split 1:2. The cells were used for transfection experiments until 13–15 passages after thawing.

Preparation of the magnetic lipoplexes:
The luciferase reporter plasmid p55pCMV-IVS-luc+, which contains the firefly luciferase cDNA under the control of the cytomegalovirus (CMV) promoter (pLuc), was amplified and purified. The GFP reporter plasmid containing the enhanced green fluorescence protein (peGFP) sequence under the control of the CMV promoter was expanded in E. coli and purified using the Qiagen plasmid purification kit.
Magnetic lipoplexes were prepared with eGFP or luciferase plasmid DNA and the DreamFect™ Gold (DF-Gold) as an enhancer at a MNPs/DF-Gold/pDNA ratio of 0.5:4:1 (iron w/v/w). For magnetofection of the adherent cells, PEI-Mag2 nanoparticles (5 µg Fe in 20 µl of water) were mixed with 40 µl of the DF-Gold, followed by the addition of 40 µl (10 µg) of the pLuc solution in a serum- and supplement-free RPMI 1640 medium. The suspension was kept at room temperature for 20 min to allow complex assembly. Then, the volume was adjusted to 2500 µl with supplement-free RPMI 1640 medium, and 25 µl of the complex solution was added to the 25000 cells in a well, resulting in a dose of 4 pg plasmid per cell.
For the magnetofection experiments involving the suspension of Jurkat T cells, the SO-Mag5 MNP suspension (25 µg Fe) was mixed with 150 µl of DF-Gold, followed by the addition of 5 µl (25 µg) of the peGFP solution and 5 µl (25 µg) of the pLuc solution in serum- and supplement-free RPMI 1640 medium for a total volume of 500 µl. The DF-Gold/pDNA lipoplexes were prepared using water instead of the MNP suspension. The complexes were applied to the Jurkat cells at a dose of 10 pg of total

plasmid per cell. The use of mixed complexes containing both the pLuc and peGFP plasmids allowed for the analysis of the protein expression and the percentage of transfected cells with the same probe.

Transfection of adherent epithelial cells with magnetic lipoplexes using different exposure patterns:

The developed rotational magnetic system was designed for the treatment of cell populations confined in detachable wells. Therefore, H441 cells were trypsinised and seeded in the Nunc BreakApart lose wells, as depicted in Figure 48, and resuspended in the cell culture medium at a density of 1.67×10^5 cells/ml. 150 µl aliquots of the obtained cell suspension were then transferred to the detachable wells of a 96-well plate, resulting in 25,000 cells per well and a confluence of approximately 40%. After incubating the cells for 24 h, 25 µl of one of the prepared magnetic transfection complexes, which contained 100 ng of plasmid, was added to each well. The plates were then exposed to a magnetic field for 15 minutes. For this, a magnetic plate (see Figure 48) was placed under the BreakApart plate with the detachable wells to induce sedimentation of the magnetic transfection complexes at the cell membrane, following the principle of classic magnetofection.

Figure 48: Top: magnetic plate used in classical magnetofection, bottom: the used detachable wells of the Nunc BreakApart strip 96-well plate, as well as their grip stoppers (blue)

The transfection experiments were conducted with respect to three patterns of exposure to the magnetic field. All wells were closed with grip stoppers and exposed to the static magnetic field of the magnetic plate for 15 minutes (reference group). The duration of the exposure to the magnetic field pulses was 30 seconds for all of the samples in the experimental groups.

In the first pattern, the cells were exposed to a pulsating magnetic field directly after exposure to the static magnetic field, i.e., at time $t = 0$ s. The second treatment pattern included 30 s of exposure to the magnetic field pulses at $t = 0$ s and again one hour after exposure to the magnetic plate (at $t = 1$ h). In the third pattern, the cells were treated with pulses only one hour after exposure to the static magnetic field.

The patterns of the magnetic field pulses:
- Group 1: only at $t = 0$ s
- Group 2: double exposure, at $t = 0$ s and $t = 1$ h
- Group 3: only at $t = 1$ h
- Group 4: reference group (only static magnetic field applied)

Five magnetic field frequencies (2.66, 5, 10, 50, and 100 Hz) and two average magnetic field amplitudes at the location of the cells (190 and 310 mT) were assessed.

After exposure to the static and pulsed magnetic fields, all of the samples were incubated at 37°C in a humidified atmosphere containing 5% CO_2 until the reporter gene analysis was performed. Each group included triplicate samples.

The expression of luciferase was assayed by measuring the chemiluminescence intensity of the luciferin-luciferase light-producing reactions in the cell lysate (Mykhaylyk O. A., 2007). The significance of the difference between the data sets was analysed using Student's t-test.

Transfection of suspension Jurkat T cells:
To explore the effect of pulsating magnetic fields on transfection efficiency, it was important to consider not only adherent cells, but also cells in suspension. For this, Jurkat T cells were labelled with anti-CD2 MicroBeads and treated by magselectofection which is defined as the combined isolation and stable nonviral transfection of cells using magnetic fields. 400 μl of magnetic lipoplexes were allowed to seep into the Miltenyi LS cell separation column, which was subsequently positioned in a MidiMACS Separator magnet to immobilise the homogeneously distributed magnetic vectors. The magnetically labelled cells (2 million) were applied to a vector-modified column, and the column was then positioned in the MidiMACS magnet, where it remained for 30 minutes at room temperature. Subsequently, the target cells were flushed from the columns with 2 ml of cell culture medium into 15

ml test tubes by pressure-enforced elution. Additionally, non-labelled cells were transfected with DF-Gold/pDNA complexes (lipofection), and CD2-Microbead-labelled Jurkat T cells were associated with the magnetic triplexes; both processes occurred at a cell density of 1 million per millilitre without the application of the static magnetic field. For the treatment with rotational magnetic field pulses, 150 µl aliquots of the cell suspension were transferred to the detachable wells and placed into the holder. The alternating magnetic field was then applied for a duration of 120 s, and at the following pulse frequencies: 2.66, 5, 10, 20 and 50 Hz. The average magnetic field amplitude was 190 mT at the location of the cells. Reporter gene expression analysis was conducted 48 hours after magselectofection using fluorescence-activated cell sorting (FACS) analysis (based on eGFP) or a luciferase assay (Yolanda Sanchez-Antequera, 2010).

5.2.3 Results

Adherent cells:

Enhanced transfection of adherent cells induced by magnetic pulses:
In accordance with the three above mentioned treatment patterns, the adherent H441 cells were transfected and treated with pulsed magnetic fields, after which they were analysed based on the luciferase reporter expression. The pulsatile transfection approach with pulse amplitudes of 190 mT and 310 mT generated, in almost all samples, a higher luciferase expression than that of the reference group of samples, which were treated only with the static magnetic field.

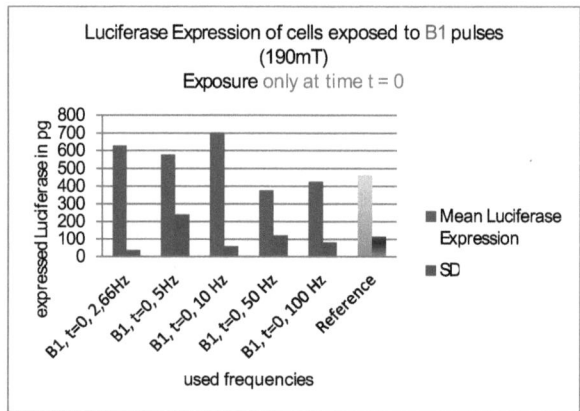

Figure 49: Higher transfection rates achieved according to first pattern (t = 0s)

Especially the treatment according to the first pattern at lower frequencies, from 2.66 to 10 Hz, led to higher transfection rates than those in the reference samples, whereas the two other pulse field patterns delivered more moderate enhancements in the transfection efficiency.

As can be further seen in Figures 49 and 50, increases in reporter gene expression of 36% and 52% at frequencies of 2.66 Hz and 10 Hz, respectively, (t=0 treatment pattern) relative to the reference data (only static magnetic field application) were observed, and fulfilled the quality of statistical significance (p=0.02 and p=0.004, respectively). A slightly negative effect on the transfection rate was observed for the higher frequencies of 50 and 100 Hz when the cells were exposed to the pulsating field immediately after the basic treatment, leading to a decrease in the transfection efficiency.

An increase in the transfection rates, yet to different extents, was reached by the stronger magnetic pulses of 310 mT. It is worthy to note that the lower frequencies persisted in being the frequencies showing the higher effects. The relative increase of 27% for f = 2.66 Hz after double exposure at t = 0 and t = 1 h and the augmentation of 23% for f = 10 Hz after single exposure at t=1 h relative to the transfection rate of the reference were statistically significant (p=0.015 and p=0.022, respectively). The exposure pattern that was limited to a single treatment after one hour with magnetic pulses of both 190 mT and 310 mT resulted in a moderate increase for almost every implemented frequency (Ch. Dahmani, 2012).

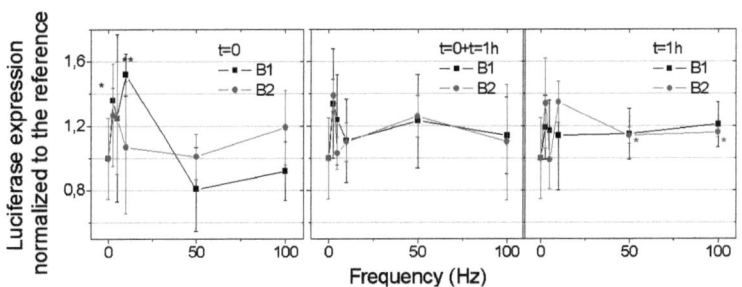

Figure 50: Luciferase expression in adherent H441 human lung epithelial cells after magnetofection and 30s exposure to pulsatile magnetic fields following the three treatment patterns (Ch. Dahmani, 2012)

Evaluation of cell viability:

To assess the cell viability after the exposure and treatment with magnetic fields, the magnetically transfected cells were washed with PBS after 72 h of incubation with the complexes. An MTT-based assay was used and the cells were further incubated for 2 h in 100 µl of 1 mg/ml MTT prepared in PBS with 5 mg/ml glucose in a well plate. Afterwards, 100 µl of a solubilisation solution containing 10% Triton X-100 in anhydrous isopropanol containing 0.1 N HCl was added to each well to dissolve the formazan. The optical density was measured at 590 nm, and the cell viability was expressed as the respiration activity normalised to the reference data for the untreated cells. Cytotoxicity was determined by measuring the total protein content with BioRad assay (Ch. Dahmani, 2012). The results in terms of percent of viable cells were quantified compared to the untreated cells and are displayed in Figure 51.

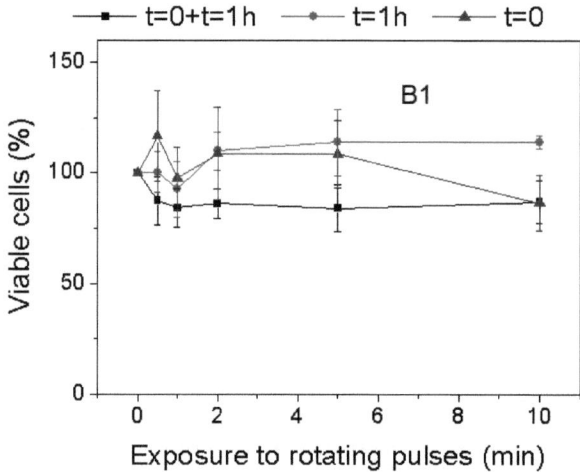

Figure 51: Percentage of viable adherent H441 cells versus duration of exposure to magnetic fields pulses (Ch. Dahmani, 2012)

The results obtained clearly proved that the percentage of viable adherent cells after pulse application remained above 75 % for the three exposure patterns, whereas the double exposure was generated the lowest viability results.

Furthermore, no statistically significant differences could be observed between the effects of the pulsed magnetic fields with amplitudes of 190 mT and 310 mT. The greatest improvement in the transfection efficiency of adherent H441 cells was observed for the lowest frequency of 2.66 Hz for all pulsed MF treatment patterns. Table 9 shows the mean change in transfection efficiency (% of reference) when using pulsating magnetic fields for all five used frequencies.

Frequency (in Hz)	190 mT	310 mT
2.66	34.9 %	32.8 %
5	24.1 %	13.6 %
10	31.2 %	8.4 %
50	2.2 %	13.5 %
100	2.8 %	14.5 %

Table 8: Mean change in transfection efficiency expressed in percentage over reference (no magnetic field use), for the amplitudes 190 and 310 mT. The data are obtained as averages from the patterns $t = 0$ and $t = 0 + t = 1h$.

Jurkat T cells in suspension

Major results involving suspension Jurkat T cells were obtained through the application of alternating magnetic field pulses for a duration of 120s and at a frequency of 10 Hz. For instance, a 1.7-fold increase in transgene expression after spontaneous association of the magnetically labelled cells with the magnetic vectors was observed. Similarly, a 1.9-fold transgene expression increase after magselectofection of the magnetically labelled cells could be documented as displayed in Figure 52.

The most significant effect generated by the pulsating magnetic field was recorded at 5 and 10 Hz, and cell viability levels remained over 100 % for all tested frequencies (2.66, 5, 10, 50 and 100 Hz). In Figure 52, statistical significance is rated with a double star ** for $p < 0.01$ and a single star * for $p < 0.05$, and samples without exposure to magnetic field pulses (=Frequency 0 Hz) were used as a reference.

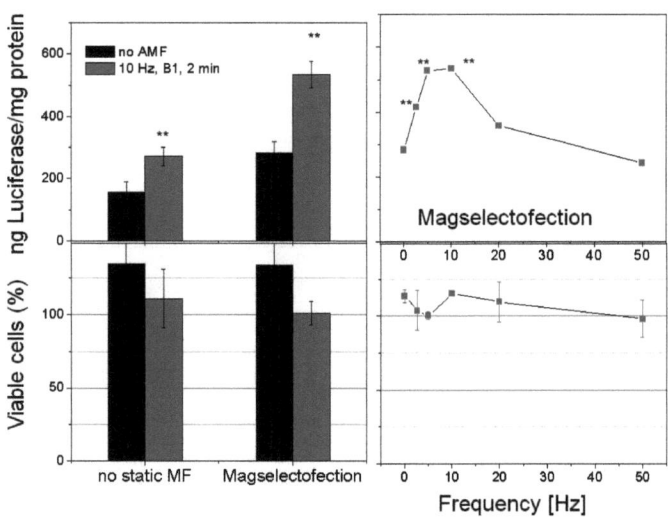

Figure 52: Left - Effect of the pulsating magnetic fields on luciferase expression and cell viability in Jurkat T-cells 48 h after transfection (pulse amplitude B1=190 mT) compared to status with no magnetic field, Right - reporter expression and cell viability as a function of the frequency of the applied alternating magnetic field after magselectofection of the cells with the complexes bearing eGFP coding plasmids (Ch. Dahmani, 2012)

Impact of the alternating magnetic field on the transfection mechanism:

After the observations made and covering adherent as well as suspension cells, evidence was gained that the increase in transfection efficiency was induced by the exposure to magnetic pulses. This resulted more specifically from the magnetic responsiveness of the magnetic DNA lipoplexes.

In comparison with classical static magnetic fields, the rotating magnetic cuboids used in the elaborated system generate defined pulses at the sites of the wells containing the magnetic vectors and the adherent as well as suspension cells, which exerts strong horizontal drag forces on the magnetic complexes.

Microscopy images, as shown in Figure 53, reveal the spatial distribution of the magnetic complexes after standard magnetofection as compared to their arrangement after the additional exposure to pulsed magnetic fields that generated lateral magnetic forces.

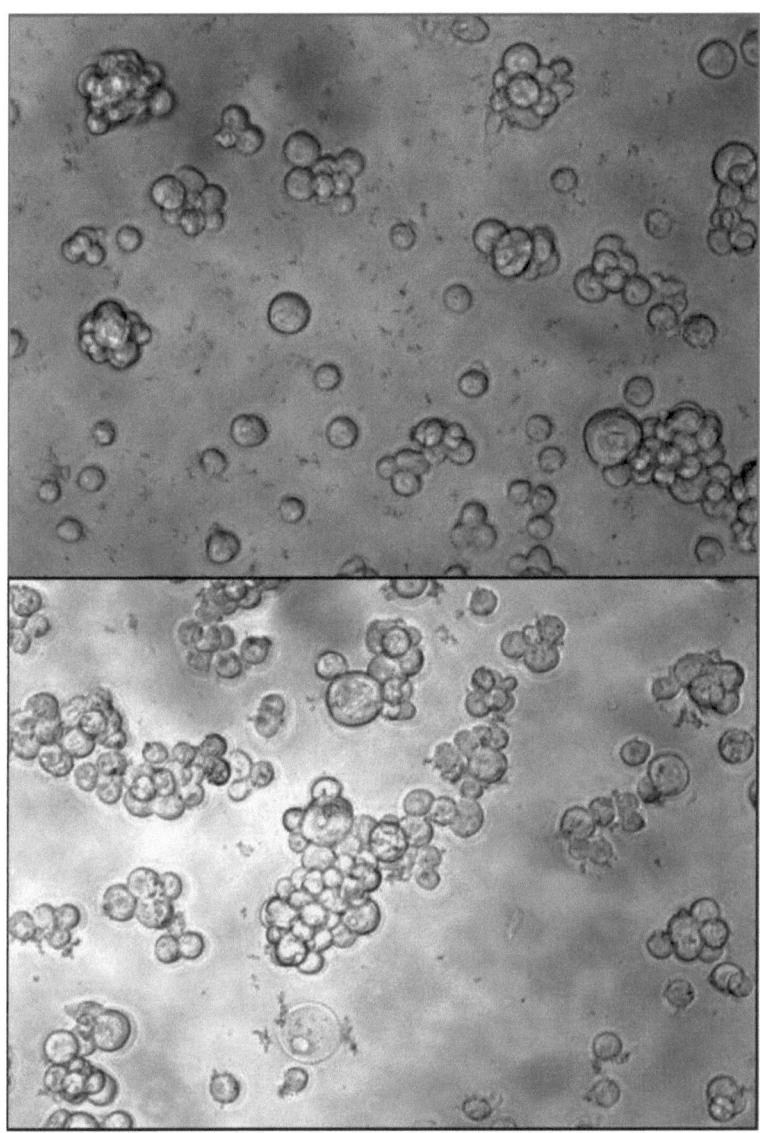

Figure 53: Comparative microscopic images of the H441 cells between (top) standard magnetofection (treatment by addition of the magnetic vectors and incubation for 20 min in the presence of the magnetic plate) and (bottom) "dynamic magnetofection" (30 s of treatment with a pulsating field at 2.66 Hz, 190 mT)

In fact, standard magnetofection implies that the MNP-containing complexes are attracted to the magnetic plate placed under the wells and therefore forced to sediment on the cell surface. Yet, a significant portion of the complexes ends up deposited at the bottom of the plate in the space between the cells and does not actively contribute to the cell internalisation and transfection process, thus considered to be "lost". In contrast, MNP/DNA-complexes exposed to the horizontal forces generated by the rotating magnets move laterally in the suspension and have an increased probability of coming into contact with a cell, being bound to the cytomembrane and being internalised into the intracellular space.

The main effect of the magnetic forces on the complexes deposited on the bottom of the well is primarily the displacement caused by the magnetic attraction, which can be described by Newton's second law of motion

$$\sum \vec{F} = m \cdot \vec{a}$$

where $\vec{F}$ describes the forces acting on the MNP/load-complexes, such as the magnetic force, hydrodynamic resistance, gravity and friction (Timothy E. Vaughan, 1996). The magnetic force is the major driving force in this consideration and increases with increases in the magnetic moment and gradient of the field. The latter reaches values of more than 50 T/m in the vicinity of the cells, as shown in Figure 47. A movement of the MNP-DNA complexes in the culture medium parallel to the bottom of the well can be actuated by the field gradients and increases the probability for a complex to collide with a cell membrane. As schematically illustrated in Figure 54, the direction of the applied magnetic force plays a major role in facilitating the movement of dragged complexes.

Beyond single magnetic nanoparticles, also entire cells loaded with magnetic complexes on their surfaces or encapsulated in their intracellular space similarly experience magnetic forces and can be dragged and moved in the suspension. This movement of the cells facilitates the formation of cell "islands" and agglomerations, as observed in the microscopic images of Figure 53.

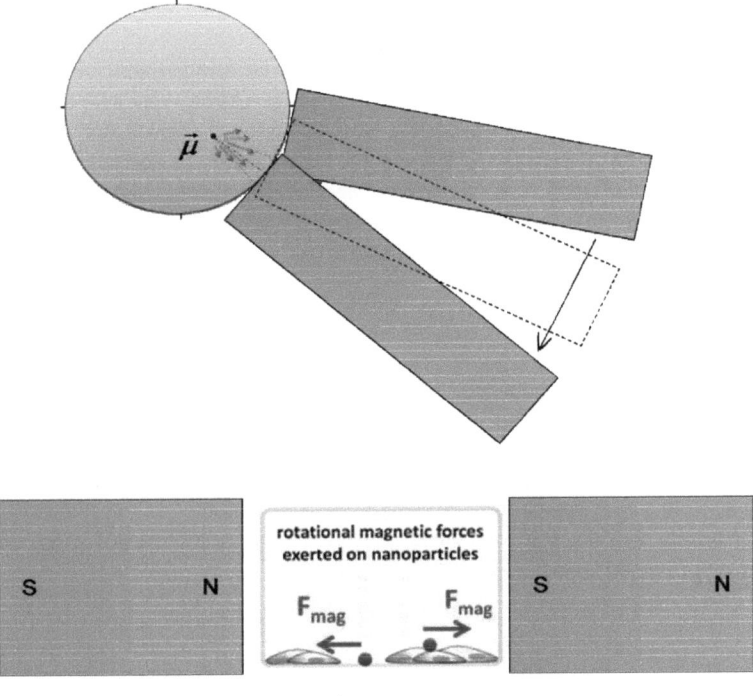

Figure 54: Schematics of the rotating magnetic forces acting on the magnetic vectors/nanoparticles in the well with cells. The applied magnetic fields were generated by rotating magnet cuboids (Ch. Dahmani, 2012)

In the literature, it was proposed by Fouriki et al. that a mechanism of oscillating nanomagnetic transfection can be generated through the mechanical stimulation of endocytosis when "an oscillating magnet array below the surface...pulls the particle into contact with the cell membrane and drags the particles from side-to-side across the cells" (A. Fouriki, 2010). Fully aligned with this explanation, it has been indeed reported, that mechanical stimuli can have effects on cellular membrane trafficking processes, including endo- and exocytosis (Apodaca, 2002).

Beyond the capacity of pulsed magnetic fields to drag the magnetic vectors to the cells, they also enable the creation of transient cell membrane openings. These so-called pores are created either by direct impact on the membrane or by the activation of a particular vibration mechanism caused by the resulting repeated attraction and repulsion action on the nanoparticles (Timothy E. Vaughan, 1996), driven by the applied magnetic forces.

When considering an energetic approach, it can be demonstrated that during sufficiently long magnetic field pulses (10^{-5} s < t_{pulse} < 10^{-1} s), very little energy is dissipated, and the rotational energy from the membrane-associated magnetic particles could be deposited temporarily into the membrane as elastic energy, as a prelude to pore creation, as suggested by Vaughan et al (Timothy E. Vaughan, 1996). For the pores of radius r_{pore} = 1 nm and membrane edge energy density $\gamma = 2 \times 10^{-11}$ J m^{-1}, the pore creation energy can be estimated to be

$$W_{pore} \approx 2\pi\gamma\, r_{pore} \approx 30\ kT \tag{39}$$

Consequently, to generate pores of radius r_{pore} = 10 and 20 nm, the energy values needed are 300 and 600 kT, respectively. It can furthermore be assumed that the complexes used in the transfection treatment, having a diameter of 1.1 μm and containing a great number of nanoparticles, are stable structures generating as whole complexes rotational elastic energies.

It follows from this that the rotational elastic energy provided by the pulses and magnetic transfection complexes is calculated to

$$W_{rot,\ elas} \approx \mu^2 \cdot B^2/2\kappa \approx 784\ kT \tag{40}$$

where $\mu = 9.3 \times 10^{-15}$ A·m² is the evaluated magnetic moment of the complexes, B = 190 mT is the magnitude of the pulse and $\kappa = 0.968 \times 10^{-18}$ Nm is the torsional spring constant of the cell membrane.

This deployed energy is sufficient to invoke a cell membrane disruption and create pores that would allow the internalization of fragments of the complexes, specifically single or gathered particles with a size of about 20 nm or parts of the borne substances associated with the NPs and present in the complexes. Thus, the magnitude of the magnetic field pulse (at low frequency and with a magnetic flux density of 190 mT at the cell layer) and the magnetic moments of the magnetic gene delivery vectors applied in the transfection experiments satisfy the energetic constraints for creating cell membrane openings. These openings may be responsible for the increased transport of material into and out of the cell and for the observed increase in transfection efficiency compared with classical magnetofection using static magnetic fields (Ch. Dahmani, 2012).

The results presented in this chapter, featuring increases in reporter luciferase gene expression in adherent human lung epithelial cells and suspension Jurkat cells, endorse the model of cell actuation mechanisms. For instance, magnetic actuation or "nano-magnetic actuation" with alternating magnetic fields clearly enables the modulation of cellular uptake and trafficking of the magnetic delivery vectors.

The enhancement of transfection efficiency through magnetic pulses comes as the ideal complement to the Magnetic Drug Targeting approach. The application of magnetic nanoparticles to the organism, followed by their guidance to and retention at target site, complemented by enhanced uptake and successful internalization of the carried bioactive load into the concerned tissue constitute the optimized scenario as for the functioning of this novel therapy method.

6 Discussion and conclusions

Gradually, targeted delivery of therapeutic substances by means of magnetically guided nanoparticles is proving to be a viable and promising new technology that still is in its ramp up phase. This thesis contributes to the advances achieved on this field through the exploration of a number of aspects of Magnetic Drug Targeting and through the development and validation of new solutions to implement it in-vitro and in-vivo.

In this work, evidence has been provided, that appropriately generated magnetic fields can efficiently retain magnetic nanoparticles in the blood vessels. Results obtained in experiments on the dorsal skin-fold chamber model demonstrated the ability to concentrate therapeutic agents at a target site in the vasculature of an organism. The retention was reported in veins, but more importantly in arteries. The site directed delivery of therapeutic agents, including genes for gene therapy, is therefore a feasible technique and can be considered as a necessary complement of a systemic drug application, the expected main advantages being a significant reduction of side-effects at non-target areas and the better control of injected doses.

However, the reported achievements were essentially based on the unprecedented proximity that has been reached between the blood vessels in the skin and the magnetic field source. A similar condition is yet relatively hard to fulfill under normal circumstances. When projecting the MDT as a future validated clinical application, the involved magnetic systems have to be advanced to a level that allows for strong field gradients at larger distances from the magnet tip. Tumors located deep within the body and not allowing the introduction of the magnetic system to their close surrounding would represent a dead zone for MDT application. Therefore, it is of utmost importance to further concentrate on the development of more sophisticated and stronger magnetic systems. A potential solution to overcome this problem can be for example the integration of field distortion and enhancement devices into a strong magnetic field generated by a magnetic resonance scanner, to produce high field gradients even at deeper organs in the body. These elements might also be intracorporal implants. Beyond this aspect, special care has to be taken regarding the emboli produced during nanoparticle retention. Especially in the case of capillaries, a blocking of the new blood supply could be the direct consequence, leading to damages of the organs or tissues irrigated by these vessels. On the other hand, selectively and reversibly embolizing parts of an organ's vasculature can certainly be an efficient way to isolate and treat it. Chemoembolization is an example of such an approach.

In the part aiming at exploring Lung Drug Targeting, simulations have shown that a significant concentration and targeting of aerosols bearing magnetic nanoparticles in the lungs could be realized, provided intubation and magnetic field triggering are taken into consideration and adequately parameterized. Learnings generated through the in-vivo exams on the big animal model were integrated in the definition of the simulation setup and the CFD computations. For instance, intubation has been modeled and the airflow in the upper respiratory pathways has been simulated accordingly. It has been demonstrated that the assumption of a continuous laminar airflow has to be corrected through the consideration of a number of eddies especially in the transitional region between the main windpipe (trachea) and the bronchi. This is due to the high velocity of the inhaled air coming out of the intubation inlet. The same high velocity effect also causes a significant deposition of aerosols at the main bifurcation region (carina ridge). The intubation is on the one hand useful, as it enables avoiding unwanted aerosol loss in the mouth and larynx level, on the other hand, it has to be appropriately adjusted, for example through prolonging the used tubes to overcome the main bifurcation and eventually reach deeper compartments of the lungs. Here, it should be pointed out, that a requisite for deep intubation is the general anesthesia of the patient, which might encounter "acceptance" issues in clinical routine.

The other major finding presented in this work is the impact of magnetic fields on magnetically active aerosols during Lung Drug Targeting. As a matter of fact, it has been demonstrated that a significant agglomeration of nanomagnetosols in a region of the lungs exposed to a high magnetic field gradient can be reproducibly achieved. An optimal constellation of the studied parameters has been identified, mainly characterized by a relatively low air flow rate of 15 l/min, and the reported influence of time variant magnetic field activation has proven that it is indispensable to synchronize the magnet field with the breathing process for a selective trapping of magnetic aerosols. The embedding of real magnetic field data into the computational fluid dynamics simulation within a realistic geometry of the lungs has therefore led to a better understanding of the Lung Drug Targeting approach and the optimal intubation conditions. The limitation of these results remains though dependent on the simplified model and the assumptions related to the simulation. A consideration of all respiratory pathways is not feasible with the anatomical information available at the moment and given the computational capacity at hand. Despite the numerous hurdles encountered during implementation, simulating the LDT process can be a considerable contribution to the fine tuning of animal experimentations, mainly leading to the reduction of the number of those.

The further significant achievement presented in this thesis relies in the reported enhancement of cell transfection efficiency by means of magnetically guided nanoparticles. For instance, a magnetic system capable of generating defined magnetic pulses was constructed and used to investigate the effect of pulsating magnetic fields on the transfection process. The approach demonstrated increases in the transfection efficiency of up to 52% in adherent H441 cells and 70-90% in the reporter gene expression in Jurkat cells, compared with the use of a static magnetic field (standard magnetofection technique). Accordingly, magnetic fields can be used in every step of the nanoparticle based site-directed drug delivery technique, from guiding the nanocarriers within the organism, through their retention and agglomeration, to their internalization into the target cells and the deployment of a therapeutic effect. One of the key challenges facing these results is the transition to pre-clinical and clinical consideration. The pattern of exposure to the magnetic pulses has to be reproducible in the context of a human application of MDT. This means that novel magnetic systems, modified and adapted to suit an upcoming clinical use, have to be developed to generate the identified pulses.

With respect to the monitoring and evaluation of the success of an MDT application, current approaches have to be revised and complemented. First, there is still no available solution to monitor -in real time- the nanoparticles' advancement or distribution in the body during treatment, even if they are tagged with a specific moiety, such as fluorescent or radioactive substances. Given the concentrations involved in MDT, it is expected to encounter major difficulties, when trying to visualize the magnetic nanoparticles under X-rays or magnetic resonance imaging (MRI). Moreover, a post-mortem evaluation of the iron-content in the targeted organs or tissues is not applicable to a clinical use case. Therefore, special approaches have to be developed to tackle the visualization issue in the magnetic drug targeting therapy.

7 Outlook

One of the most pertinent fields, where Magnetic Drug Targeting can have a significant contribution to the improvement of medicine, is undoubtedly cancer treatment.

7 million deaths per year in the world can be attributed to cancer, representing approximately 12.5% of total deaths. It is believed that any person who lives long enough will eventually get cancer.

However, strong signals emitted by the achievements realized in the last decades in cancer therapy suggest that medicine is capable of also surmounting this challenge.

In fact, many of today's established treatments were still in their experimental phase only a decade ago, so current clinical studies of new cancer therapy methods may translate into tomorrow's treatment for most of the known tumor types.

In analogy to traditional vaccines that have successfully prevented infectious diseases that were thought to be invincible, scientists continue to investigate ways to better understand, diagnose, and treat tumors.

More recently, the great progress made in surgery, chemotherapy and radiation therapy, has led to the reduction of mortality caused by cancer. Gradually, traditional treatments are being supplemented with newer methods including chemoprevention and cancer vaccines. A further, particularly promising approach, are Nanotechnology-based efficient drug delivery systems. Here, the intrinsic, size-based proximity between nanostructures and the elementary constituents of the organism (cells and cell organelles) suggests a better interaction and an enhanced therapeutic effect. The use of engineered nano-devices and nanostructures, at the single cell level, is expected to ultimately change the way medical benefit is achieved.

In accordance with these outlooks, Magnetic Drug Targeting essentially enhances the specificity and selectiveness of chemotherapy. It offers a new way to carry drugs specifically to tumors, which would help resurrect promising drugs that failed in clinical trials because they were cleared from the bloodstream before they could reach their intended targets, or had to be given in doses high enough to produce toxic side effects.

Furthermore, MDT enables new treatment approaches, such as gene therapy. In future projections of medical treatments, the underlying principle of healing is expected to evolve from the conventional destroying or eliminating of a malignant tissue to its preservation, repair and improvement. Here too, MDT presents an enormous potential. For instance, it can be applied to advance immunotherapies that aim at enhancing the body's overall immune response to recognize and fight cancer cells. Mobilization of the human body itself against cancer may, in fact, provide a very attractive alternative or complement to conventional chemotherapies and radiation therapy.

This is but one of the many examples where nanotechnology will bring about the development of the so-called regenerative medicine, towards a cell-by-cell treatment and monitoring capability.

„Der Geist der Medizin ist leicht zu fassen! Ihr durchstudiert die groß - und kleine Welt, um es am Ende gehn zu lassen, wie's Gott gefällt." (Mephistopheles)

Johann Wolfgang von Goethe
(28.08.1749 - 22.03.1832)

List of Tables

Table 1: Comparison of the major established methods of magnetic nanoparticle synthesis (Song Ge, 2009). 12

Table 3: Selected materials and the corresponding critical sizes below which they display a single-domain structure (An-Hui Lu, 2007) 23

Table 4: Relevant properties of the PEI-Mag2 magnetic nanoparticles 25

Table 5: Different removal processes of NPs (Arruebo Manuel, 2007) 34

Table 6: Parameters for the coil simulation 47

Table 7: Parameters varied to study particle deposition under magnetic force application in a lung model geometry (Issa, 2010) 67

Table 8: Lung modeling data according to Horsfield's Model 1 68

Table 9: Mean change in transfection efficiency expressed in percentage over reference (no magnetic field use), for the amplitudes 190 and 310 mT. The data are obtained as averages from the patterns $t = 0$ and $t = 0 + t = 1h$. 104

List of Figures

Figure 1: Composition of a standard nanoparticle system for biomedical applications (adapted from (Salata 2004)) 11

Figure 2: Relationship between the percentage of surface atoms in a particle and its size (Halim, 2008) 22

Figure 3: Minimizing magnetostatic energy stored in the external magnetic field of a material sample through the creation of internal magnetic domains. The magnetic field lines extend in loops in opposite directions through each domain, which reduces the field outside the material (Dutz, 2008). 23

Figure 4: Hysteresis loops for ferromagnetic and superparamagnetic nanoparticles, in comparison with para- and diamagnetic behavior. 24

Figure 5: Measurement of the magnetic responsiveness of the magnetically labeled cells and the magnetic nanoparticles 26

Figure 6: Measurement setup to determine the total magnetic moment of a given nanoparticle charge (left), Nanoparticle suspension exposed to magnetic force for magnetic moment measurement (right) 29

Figure 7: Nonlinear increase of the total magnetic moment of a sample with augmented sample mass as proof of the Cluster effect – here measured for FeSi magnetic nanoparticles 29

Figure 8: SEM images of spindle shaped nanoparticles after spin coating on a silicon wafer and sputtering (titan + gold) - top-left: a fluidMAG/SP-D particle/ top-right: two adjacent fluidMAG/SP-D particles/ bottom: the sample comprising multiple nanoparticles whereas their particular shape was lost through processing for SEM imaging. 31

Figure 9: Fate of nanoparticles after injection for cancer therapy (Challa S.S.R. Kumar, 2005) 34

Figure 10: Dependence of the blood residence time of NPs on their size (Arruebo Manuel, 2007) 35

Figure 11: Left - Schematic illustration of an NDA-MNP-Enhancer complex comprising 110000 particles and 47000 plasmid copies as described and used in (Mykhaylyk, Steingötter, Perea, Aigner, Botnar, & Plank, 2009), Right – a conglomerate of magnetic nanoparticles (FluidMAG-D) from the company Chemicell® as visualized under scanning electron microscopy 37

Figure 12: Schematic illustration of major forces acting on a magnetic nanoparticle navigated in blood vessels 40

Figure 13: Front shapes of a plug flow (top) and a parabolic flow (bottom) in large and small blood vessels (Leach, 2003).. 40

Figure 14: Model geometry of the designed magnet in 3D visualization 47

Figure 15: Simulated magnetic flux density of the designed electromagnet (top), and plot of arrows symbolizing the magnetic field lines surrounding the magnet tip (Hoke, 2008) (I. Hoke, 2008) 48

Figure 16: Plot of the magnetic field density and gradient under with distance from the magnetic tip 49

Figure 17: construction plans for the electromagnet, with a modular assembly design and an adaptable tip arrangement.. 50

Figure 18: Constructed experimental electromagnet, with its exchangeable tip system (bottom) and after assembly (top). The top left figure shows the magnet after galvanization and featuring a temperature measurement strip. 51

Figure 19: The dorsal skin-fold chamber model, where part of the dorsal skin of a laboratory mouse is stretched along the window of a fixing device, allowing the visualization of the vasculature as well as the exposure to strong magnetic fields 53

Figure 20: Flipped arrangement of the magnetic system with the upwardly oriented tip, allowing for better accessibility to the vessels of the skin-fold chamber of the mouse 54

Figure 21: Fixation device including skin-fold chamber mounted on top of the magnet tip... 55

Figure 22: Zeiss Microscope placed few millimeters above the skin-fold chamber that exposes the confined part of the dorsal vasculature to a high magnetic field (1039 mT). The configuration features fluorescence imaging (right) and enables real time assessment of the retention efficiency on a display (left).................. 56

Figure 23: Fluorescence tracking under microscopy of microbubble agglomerations in the arteries of a mouse (dorsal skin). From top to bottom, the pictures show the gradual formation of the agglomeration, including at bifurcation, under magnetic field influence... 58

Figure 24: Results of site specificity of gene delivery mediated by magnetic microbubbles as achieved in the skin-fold chamber vasculature of a mouse

(Hanna Mannell, 2012). The expression of the applied plasmid DNA was detected in a significantly differentiated way in the targeted site. 59

Figure 25: Illustration of the main constituents of the human lung comprising the trachea, bronchi, bronchial tree and ending up at the alveolar level (National Cancer Institute, 2012) 63

Figure 26: Denomination of the upper respiratory pathways and main bifurcations, as derived from Horsfield's Model 1 (K. Horsfield, 1971) 68

Figure 27: Designed geometry (top left) and mesh (top right and bottom) of the upper pulmonary airways, based on the Horsfield's lung data and the physiologically realistic bifurcation computation approach (Gläßner, 2008) 69

Figure 28: Geometry (top) and Mesh (bottom) of the simulated Siemens magnet... 71

Figure 29: Magnetic flux density in a layer 2 mm under magnet tip. 72

Figure 30: Magnetic field gradient (bottom) in a layer 2 mm under magnet tip. 73

Figure 31: Measured magnetic field data (flux density and field gradient) – Siemens magnet (C. Alexiou, 2006). 74

Figure 32: Simulated magnetic flux density (top) and simulated magnetic field gradient (bottom) for distances given in mm under magnet tip. 75

Figure 33: Illustration of extraction layers where the magnetic field data are plotted under the magnet tip. 76

Figure 34: Retention effect on MNP loaded aerosols subjected to a magnetic field gradient positioned over the left side of the lung (Issa, 2010). 78

Figure 35: Selected results of particle behavior and sedimentation in the proposed lung model under magnetic field activation (Issa, 2010) 79

Figure 36: Effect of intubation on aerosol deposition rate, with varying aerosol diameter (blue line: no intubation, green and red lines using tubes of different extremity shapes) (Issa, 2010) 80

Figure 37: Experimental setup for the preclinical Lung Drug Targeting studies conducted on the big animal model (pig) and featuring the positioning of the electromagnet tip. 82

Figure 38: Distribution of radioactively marked nanoparticles in the lungs of a pig after LDT application. 83

Figure 39: Reconstruction of the upper tracheo-bronchial tree of three different pigs treated in the pre-clinical evaluation and exhibiting a first minor bifurcation before the main bronchi 84

Figure 40: Reconstruction of the upper bronchial tree of an experimentation pig including the tube used for ventilation. The inclination of the tube explains the unusually increased aerosol deposition in the upper trachea. 84

Figure 41: Time schedule for magnetic field activation in an optimized, breath-synchronous Lung Drug Targeting procedure (Dahmani Ch., 2008).................. 86

Figure 42: Natural mechanisms of cellular drug uptake into tumor tissue, as illustrated in (Challa S.S.R. Kumar, 2005) 89

Figure 43: Illustration of the Magnetofection technique, consisting in the nucleic acid delivery to target cells guided and enhanced by the action of magnetic gradient fields on gene vectors associated with magnetic particles. 90

Figure 44: Magnetic flux density distribution (in T) and magnetic field lines of the simulated magnet... 93

Figure 45: Magnetic flux density distribution (in T) and magnetic field lines of the simulated magnet dispositions for a magnetization along depth of rod, parallel (C), anti-parallel (D) 94

Figure 46: Top: the constructed rotating magnetic system with brushless motor, bottom: well holding disc (right) and multilevel well holder (left) for magnetic exposure of cells.. 96

Figure 47: Magnetic flux density and magnetic field gradient at a distance $r = 7.5$ mm from the system center and a frequency $f = 10$ Hz.. 97

Figure 48: Top: magnetic plate used in classical magnetofection, bottom: the used detachable wells of the Nunc BreakApart strip 96-well plate, as well as their grip stoppers (blue) 99

Figure 49: Higher transfection rates achieved according to first pattern ($t = 0$s) 101

Figure 50: Luciferase expression in adherent H441 human lung epithelial cells after magnetofection and 30s exposure to pulsatile magnetic fields following the three treatment patterns (Ch. Dahmani, 2012)... 102

Figure 51: Percentage of viable adherent H441 cells versus duration of exposure to magnetic fields pulses (Ch. Dahmani, 2012)... 103

Figure 52: Left - Effect of the pulsating magnetic fields on luciferase expression and cell viability in Jurkat T-cells 48 h after transfection (pulse amplitude $B_1=190$ mT) compared to status with no magnetic field, Right - reporter expression and cell viability as a function of the frequency of the applied alternating magnetic field after magselectofection of the cells with the complexes bearing eGFP coding plasmids (Ch. Dahmani, 2012) 105

Figure 53: Comparative microscopic images of the H441 cells between (top) standard magnetofection (treatment by addition of the magnetic vectors and incubation for 20 min in the presence of the magnetic plate) and (bottom) "dynamic magnetofection" (30 s of treatment with a pulsating field at 2.66 Hz, 190 mT) 106

Figure 54: Schematics of the rotating magnetic forces acting on the magnetic vectors/nanoparticles in the well with cells. The applied magnetic fields were generated by rotating magnet cuboids (Ch. Dahmani, 2012) 108

List of References

A. Fouriki, N. F. (9. July 2010). Evaluation of the magnetic field requirements for nanomagnetic gene transfection. Nano Reviews, S. 1-5.

ALF LAMPRECHT, N. U.-M. (31. July 2001). Biodegradable Nanoparticles for Targeted Drug Delivery in Treatment of Inflammatory Bowel Disease. THE JOURNAL OF PHARMACOLOGY AND EXPERIMENTAL THERAPEUTICS, S. 775–781.

Andreas Lübbe, C. A. (13. Dezember 2000). Clinical Applications of Magnetic Drug Targeting. Journal of Surgical Research, S. 200-206.

An-Hui Lu, E. L. (46 2007). Magnetic Nanoparticles: Synthesis, Protection, Functionalization, and Application. Angewandte Chemie, S. 1222-1244.

Apodaca, G. (282(2) 2002). Modulation of membrane traffic by mechanical stimuli. Am J Physiol Renal Physiol, S. 179-190.

April M. Chow, K. W. (63 2010). Enhancement of Gas-Filled Microbubble R2 by Iron Oxide Nanoparticles for MRI. Magnetic Resonance in Medicine, S. 224-229.

Arruebo Manuel, R. F.-P. (June 2007). Magnetic nanoparticles for drug delivery. Nanotoday, S. 22-32.

AuroVist. (January 2009). AuroVist™ Gold Nanoparticle X-ray Contrast Agent. Product Information and Instructions Document, S. 1-6.

Benyus, J. M. (1997). Biomimicry: Innovation Inspired by Nature. New York: Harper Collins Publishers.

Bernhard Gleich, N. H. (Vol. 6. March 2007). Design and Evaluation of Magnetic Fields for Nanoparticle Drug Targeting in Cancer. IEEE Transactions on Nanotechnology, S. 164-170.

Bitterle, E. (2004). Oxidative und inflammatorische Mechanismen von Targetzellen des Alveolarepithels nach Exposition mit ultrafeinen Aerosolpartikeln an der Luft-Medium-Grenzschicht. München: Verlag Dr. Hut.

Brower, V. (4. January 2006). Is Nanotechnology Ready for Primetime? Journal of the National Cancer Institute, S. 9-11.

C. Alexiou, D. D. (Vol. 16, No. 2. June 2006). A High Field Gradient Magnet for Magnetic Drug Targeting. IEEE Transactions on Applied Superconductivity, S. 1527-1530.

C. Bergemann, D. M.-S. (1999). Magnetic ion-exchange nano- and microparticles for medical biochemical and molecular biological applications. Journal of Magnetism and Magnetic Materials, S. 45-52.

C. Damyanov, M. R. (24. February 2009). Low dose chemotherapy in combination with insulin for the treatment of advanced metastatic tumors. Preliminary experience. Journal of BUON, The official journal of the Balkan Union of Oncology, S. 711-715.

C. Wilhelm, F. G.-C. (9. Februar 2002). Magnetophoresis and ferromagnetic resonance of magnetically labelled cells. European Biophysics Journal, S. 118 - 125.

Cai X., Y. F. (2(1) 2012). Applications of Magnetic Microbubbles for Theranostics. Theranostics, S. 103-112.

Ch. Dahmani, O. M.-G. (28. 12 2012). Rotational magnetic pulses enhance the magnetofection efficiency in vitro in adherent and suspension cells. Journal of Magnetism and Magnetic Materials, S. 163–171.

Challa S.S.R. Kumar, J. H. (2005). Nanofabrication: Towards Biomedical Applications. Weinheim, Germany: Wiley-VCH.

Chao-Bin Chen, J.-Y. C.-C. (Vol. 9 2009). Fast Transfection of Mammalian Cells Using Superparamagnetic Nanoparticles Under Strong Magnetic Field. Journal of Nanoscience and Nanotechnology, S. 2651-2659.

Chemicell GmbH. (2009). Magnetofection TM 2.3: The new gene transfection technology. Berlin.

Chris Gunter, R. D. (Vol 420 2002). The mouse genome. Nature, S. 509.

Christoph Alexiou, W. A. (1. December 2000). Locoregional Cancer Treatment with Magnetic Drug Targeting. Cancer Research, S. 6641-6648.

Claire Wilhelm, F. G. (2008). Universal cell labelling with anionic magnetic nanoparticles. Biomaterials, S. 3161 - 3174.

Comsol Multiphysics, C. (2009). Modeling Guide.

Dahmani Ch., G. S. (Volume 22. November 2008). Breath Synchronous Magnetic Drug Targeting in the Lungs. Proceedings of the 4th European Conference of the International Federation for Medical and Biological Engineering.

Dames, P. (22. Juli 2007). Targeted delivery of magnetic aerosol droplets to the lung. Nature Nanotechnology.

Dialechti Vlaskou, O. M. (23. November 2010). Magnetic and Acoustically Active Lipospheres for Magnetically Targeted Nucleic Acid Delivery. Advanced Functional Materials, S. 3881–3894.

Dutz, S. (2008). Nanopartikel in der Medizin. Hamburg: Verlag Dr. Kovač.

E. P. Furlani, K. C. (17. June 2008). Nanoscale magnetic biotransport with application to magnetofection. Physical Review, S. 061914-(1) to 061914-(8).

Eberhard Kallenbach, R. E. (2003). Elektromagnete. Grundlagen, Berechnung, Konstruktion, Anwendung. Vieweg+Teubner Verlag.

EurekAlert. (12 Oct 2012). Scientists discover that shape matters in DNA nanoparticle therapy. http://www.eurekalert.org/pub_releases/2012-10/nu-sdt101012.php#: Northwestern University.

F. Scherer, M. A. (volume 9. January 2002). Magnetofection: enhancing and targeting gene delivery by magnetic force in vitro and in vivo. Gene Therapy, S. 102-109.

Feynman, R. (Vol. 1, Nr. 1. March 1992). There's plenty of room at the bottom. Journal of Microelectromechanical Systems, S. 60-66.

Fouriki A, C. M. (12. April 2012). Efficient transfection of MG-63 osteoblasts using magnetic nanoparticles and oscillating magnetic fields. Journal of Tissue Engineering and Regenerative Medicine, S. n/a-n/a.

Franzreb, M. (2003). Magnettechnologie in der Verfahrenstechnik wässriger Medien (Professorial Dissertation). Karlsruhe.

Füssl, F. (1997). Darstellung und Charakterisierung superparamagnetischer Nanopartikel sowie deren Modifizierung für den Einsatz in der Hyperthermie-Therapie. Aachen: trans-aix-press.

G. Scheuch, M. J. (58 2006). Clinical perspectives on pulmonary systemic and macromolecular delivery. Advanced Drug Delivery Reviews, S. 996 – 1008.

Gazeau, C. W. (1. May 2008). Universal cell labelling with anionic magnetic nanoparticles. Biomaterials, S. 3161-3174.

Gläßner, J. (2008). Untersuchung des Partikelverhaltens in den Atemwegen durch Stimulation. Technische Universität München: Master Thesis Report.

Gleich, B. (2007). Aktiver Wirkstofftransport mit magnetischen Feldern. München.

Gonda, I. (5 1990). Major issues and future prospects in the delivery of therapeutic and diagnostic agents to the respiratory tract. Advanced Drug Delivery, S. 1-9.

Götz, S. (2008). Atem-getriggerte Magnetsteuerung für das Lungen-Drug-Targeting. Technische Universität München: Master Thesis Report.

H. K. Chang, O. A. (49 1982). A model study of flow dynamics in human central airways. Part I: Axial velocity profiles. Respiration Physiology, S. 75-95.

Halim, S. C. (2008). Application of reactive and partly soluble nanomaterials. Zurich: ETH Dissertation Library.

Hanjie Wang, S. Z. (5(1) 2010). PEGlated magnetic polymeric liposome anchored with TAT for delivery of drugs across the blood-spinal cord barrier. Biomaterials, S. 65-76.

Hanna Mannell, J. P. (8 2012). Site directed gene delivery in vivo by ultrasonic destruction of magnetic nanoparticle coated microbubbles. Nanomedicine: Nanotechnology, Biology, and Medicine, S. 1309-1318.

Helmtrud I. Roach, a. N. (1. August 1999). Cell Paralysis as an Intermediate Stage in the Programmed Cell Death of Epiphyseal Chondrocytes During Development. Journal of Bone and Mineral Research, S. 1367-1378.

Hoke, I. (2008). Simulation einer Magnetanordnung zur Retention magnetischer Nanopartikel im Gehirn einer Versuchsmaus. München: Diplomarbeit an der Technischen Universität München.

Hongwei Gu, K. X. (19. January 2006). Biofunctional magnetic nanoparticles for protein separation and pathogen detection. Chemical Communications, S. 941–949.

Hongwei Gu, K. X. (19. January 2006). Biofunctional magnetic nanoparticles for protein separation and pathogen detection. Chemical Communications, S. 941-949.

Hyon Bin Na, J. H. (46 2007). Development of a T1 Contrast Agent for Magnetic Resonance Imaging Using MnO Nanoparticles. Angewandte Chemie, S. 1-6.

I. Hoke, C. D. (4-6. November 2008). Design of a High Field Gradient Electromagnet for Drug Delivery to a Mouse Brain. Proceedings of the Comsol Conference 2008.

Ikramy A. Khalil, K. K. (2006). Uptake Pathways and Subsequent Intracellular Trafficking in Nonviral Gene Delivery. Pharmacological Reviews, S. 32-45.

Issa, R. (2010). Multiparametric Analysis of the Behaviour of Magnetic Nanoparticles in the Airways (Lung Drug Targeting). Technische Universität München:

Master Thesis Report.

J. Stöhr, H. S. (2006). Magnetism, From Fundamentals to Nanoscale Dynamics. Berlin, Heidelberg: Springer.

Jain, K. K. (2008). The Handbook of Nanomedicine. Totowa, NJ, USA: Humana Press.

Jana Chomoucka, J. D. (21. January 2010). Magnetic nanoparticles and targeted drug delivering. Pharmaceutical Research, S. 144-149.

Javed Ally, B. M. (4. March 2005). Magnetic targeting of aerosol particles for cancer therapy. Journal of Magnetism and Magnetic Materials, S. 442–449.

Ji-Yao Chen, Y.-L. L.-H.-C. (39(3) 2006). Transformation of Escherichia coli mediated by magnetic nanoparticles in pulsed magnetic field. Enzyme and Microbial Technology, S. 366-370.

José I. Schwerdt, G. F. (No. 1. Vol. 12 2012). Magnetic Field-Assisted Gene Delivery: Achievements and Therapeutic Potential. Current Gene therapy, S. 116-126.

José Luis Corchero, a. A. (27. June 2007). Biomedical applications of distally controlled magnetic nanoparticles. Cell Press, S. 468-476.

K. Horsfield, G. D. (Nr. 2. August 1971). Models of the human bronchial tree. Journal of Applied Physiology, S. 207-217.

Kübler, B. (2007). Biomedizinische Technik 3: Medizinische Gerätetechnik. Institut für Biomedizinische Technik: Universität Stuttgart.

Kubo T, S. T. (January 2001). Targeted systemic chemotherapy using magnetic liposomes with incorporated adriamycin for osteosarcoma in hamsters. International Jounal of Oncology, S. 121-125.

L. W. Wattenberg, T. S. (64 2004). Chemoprevention of Cancer of the Upper Respiratory Tract of the Syrian Golden Hamster by Aerosol Administration of Difluoromethylornithine and 5-Fluorouracil. Cancer Research, S. 2347 – 2349.

Leach, J. H. (2003). Magnetic Targeted Drug Delivery. Master Thesis, Virginia Polytechnic Institute and State University.

M. Lohakan, P. J. (2007). A Computational Model of Magnetic Drug Targeting in Blood Vessel using Finite Element Method. Transactions of the 2007 Second Conference on Industrial Electronics and Applications, S. 231-234.

Manfred Ogris, D. O. (2012). Nanotechnology for Nucleic Acid Delivery: Methods and Protocols (Methods in Molecular Biology). Humana Press.

Marco A. Zarbin, C. M. (August 2010). Nanomedicine in Ophthalmology: The New Frontier. Nanomedicine In Ophtalmology, S. 144-162.

Matthias T Stephan, J. J. (15. August 2010). Therapeutic cell engineering with surface-conjugated synthetic nanoparticles. Nature Medicine.

McBain SC, G. U. (8. October 2008). Magnetic nanoparticles as gene delivery agents: enhanced transfection in the presence of oscillating magnet arrays. Nanotechnology, S. 1-5.

Mykhaylyk Olga, D. V. (311 2007). Magnetic nanoparticle formulations for DNA and siRNA delivery. Journal of Magnetism and Magnetic Materials, S. 275-281.

Mykhaylyk, O. A. (2. Septembr 2007). Generation of magnetic nonviral gene transfer agents and magnetofection in vitro. Nat Protoc, S. 2391-2411.

Mykhaylyk, O., Steingötter, A., Perea, H., Aigner, J., Botnar, R., & Plank, C. (December 2009). Nucleic Acid Delivery to Magnetically-Labeled Cells in a 2D Array and at the Luminal Surface of Cell Culture Tube and Their Detection by MRI. Journal of Biomedical Nanotechnology, S. 692-706.

Nanomagnetomedizin, V. (2009). Poster zum BMBF-Verbundprojekt Nanomagnetische Arzneistoffträger. Ulm: Bundesministerium für Bildung und Forschung (BMBF).

National Cancer Institute, U. N. (15. September 2012). Bronchi, Bronchial Tree, & Lungs. http://training.seer.cancer.gov/anatomy/respiratory/passages/bronchi.html.

Nikivorov A.I., S. R. (59 1985). Morphometric variability of the human upper bronchial tree. Respiration Physiology, S. 289-299.

O'Handley, R. C. (2000). Modern Magnetic Materials: Principles and Applications. New York: Wiley-Interscience.

Ovidiu Rotariu, N. J. (3. March 2005). Modelling magnetic carrier particle targeting in the tumor microvasculature for cancer treatment. Journal of Magnetism and Magnetic Materials, S. 639-646.

Ozer, A. Y. (2007). Alternative Applications for Drug Delivery: Nasal and Pulmonary Routes. Nanomaterials and Nanosystems for Biomedical Applications, S. 99-112.

P.A. Voltairasa, D. F. (5. February 2002). Hydrodynamics of magnetic drug targeting. Journal of Biomechanics, S. 813-821.

Pankhurst Q A, C. J. (36 2003). Applications of magnetic nanoparticles in biomedicine. Journal of Physics, S. R167-R181.

Pascal R. Leroueil, S. H. (2007). Nanoparticle Interaction with Biological Membranes: Does Nanotechnology Present a Janus Face? Accounts of Chemical Research, S. 335 - 342.

Pickard M, C. D. (5. February 2010). Enhancement of magnetic nanoparticle-mediated gene transfer to astrocytes by 'magnetofection': effects of static and oscillating fields. Nanomedicine, S. 217-232.

Plank C, A. M. (3(5) 2003). Enhancing and targeting nucleic acid delivery by magnetic force. Expert Opinion on Biological Therapy, S. 745-758.

Plank C., S. F. (2000). Magnetofection: Enhancement and localization of gene delivery with magnetic particles under influence of a magnetic fields. Journal of Gene Medicine, S. 24.

Plank, C. (September 2009). Silence the target. Nature Nanotechnology, S. 544-545.

Pothur R. Srinivas, P. B. (No. 5. Vol. 82 2002). Nanotechnology in Early Detection of Cancer. LABORATORY INVESTIGATION, S. 657-662.

R Jurgons, C. S. (8. September 2006). Drug loaded magnetic nanoparticles for cancer therapy. JOURNAL OF PHYSICS, S. 2893–2902.

Robert Sturm, W. ((19) 2009). Modellrechnungen zur Deposition nicht-sphärischer Teilchen in den oberen Luftwegen der menschlichen Lunge. Zeitschrift für Medizinische Physik, S. 38-46.

Salata, O. (30. April 2004). Applications of nanoparticles in biology and medicine. Journal of Nanobiotechnology.

Sang Jun Son, J. R. (18. March 2005). Magnetic Nanotubes for Magnetic-Field-Assisted Bioseparation, Biointeraction, and Drug Delivery. Journal of the American Chemical Society, S. 7316-7317.

Sanjeeb K. Sahoo, a. V. (24. December 2003). Nanotech approaches to drug delivery and imaging. Drug Discovery Today, S. 1112-1120.

Sarah W. Kamau Chapman, P. O.-S.-A.-F. (4. January 2008). Application of pulsed-magnetic field enhances non-viral gene delivery in primary cells from different origins. Journal of Magnetism and Magnetic Materials, S. 1517–1527.

Sarah W. Kamau, P. O.-F.-A. (15. March 2006). Enhancement of the efficiency of non-viral gene delivery by application of pulsed magnetic field. Nucleic Acids Research, S. e40.

Sharyn D. Baker, J. V. (18. December 2002). Role of body surface area in dosing of investigational anticancer agents in adults 1991-2001. Journal of the National Cancer Institute, S. 1883 - 1888.

Shin-ichi Takeda, F. M. (19. Dezember 2006). Development of magnetically targeted drug delivery system using superconducting magnet. Journal of Magnetism and Magnetic Materials, S. 367-371.

Shin-ichi Takeda, F. M. (19. December 2006). Development of magnetically targeted drug delivery system using superconducting magnet. Journal of Magnetism and Magneic Materials, S. 367-371.

Sibnath Kayal, R. V. (Vol. 10 2010). Anti-Cancer Drug Loaded Iron–Gold Core–Shell Nanoparticles (Fe@Au) for Magnetic Drug Targeting. Journal of Nanoscience and Nanotechnology, S. 1-13.

Sinha, A. (2008). CHARACTERIZING MAGNETIC PARTICLE TRANSPORT OF MICROFLUIDIC APPLICATIONS. Blacksburg, Virginia: Faculty of Virginia Polytechnic Institute and State University.

Song Ge, X. S. (6. August 2009). A Facile Hydrothermal Synthesis of Iron Oxide Nanoparticles with Tunable Magnetic Properties. J Phys Chem C Nanomater Interfaces, S. 13593–13599.

Sophie Laurent, M. M. (2(4) 2011). Superparamagnetic iron oxide nanoparticles: promises for diagnosis and treatment of cancer. International Journal of Molecular Epidemiology and Genetics, S. 367-390.

Stefan M. Götz, C. D. (9. September 2010). First Theoretic Analysis of Magnetic Drug Targeting in the Lung. IEEE Transactions on Biomedical Engineering, S. 2115-2121.

Sturm, R., & Hofman, W. (2009). Modellrechnungen zur Deposition nicht-sphärischer Teilchen in den oberen Luftwegen der menschlichen Lunge. Zeitschrift für Medizinische Physik, 19, S. 38-46.

Suchanek, A., & Kerscher, K.-J. (2007). Der Homo oeconomicus: Verfehltes Menschenbild oder leistungsfähiges Analyseinstrument? In R. Lang, & A. Schmidt, Individuum und Organisationen: Neue Trends eines organisationswissenschaftlichen Forschungsfeldes. (S. 252-275). Wiesbaden: Deutscher Universitäts-Verlag.

T Kubo, T. S. (Volume 11 - Number 1. January 2001). Targeted systemic chemotherapy using magnetic liposomes with incorporated adriamycin for osteosarcoma in hamsters. International Journal of Oncology, S. 121-126.

T. A. Connors, R. D. (1995). The Chemotherapy of Colon Cancer. European Journal of Cancer, S. 1373-1378.

T.K. Indira, P. L. (Volume 3 - Issue 3. October - December 2010). Magnetic Nanoparticles - A Review. International Journal of Pharmaceutical Sciences and Nanotechnology, S. 1035-1042.

Tarl W. Prow, J. H. (2004). Nanomedicine: nanoparticles, molecular biosensors, and targeted gene/drug delivery for combined single-cell diagnostics and therapeutics. Advanced Biomedical and Clinical Diagnostic Systems II , (S. Proc. SPIE, Vol. 5318, 1). San Jose, CA, USA.

Ted B. Martonen, Z. Z. (Volume 35 2001). Three-Dimensional Computer Modeling of the Human Upper Respiratory Tract. Cell Biochemistry and Biophysics, S. 255-261.

Thomas Weyh, N. S. (2004). Control of drug-carrying magnetobeads by magnetic gradient-fields. Transactions of the 4th IEEE Conference on Nanotechnology, S. 477-479.

Timothy E. Vaughan, J. C. (August 1996). Energetic Constraints on the Creation of Cell Membrane Pores by Magnetic Particles. Biophysical Journal, S. 616-622.

Torchilin, V. (2006). Nanoparticulates as drug carriers. London: Imperial College Press.

Tristan Montier, T. B.-A.-J. (8 2008). Progress in Cationic Lipid-Mediated Gene Transfection: A Series of Bio- Inspired Lipids as an Example. Current Gene Therapy, S. 296-312.

Ushka K. Veeramachaneni, R. L. (2007). Magnetic Particle Motion in a Gradient Field. Proceedings of the Comsol Conference Boston 2007, S. 1-5.

Vainauska D, K. S. (48(6) 2012). A novel approach for nucleic acid delivery into cancer cells. Medicina (Kaunas), S. 324-329.

W. E. Walker, J. a. ((Nr 49) 1968). Lysosome Response of Virus Infected Serum and Serum-free Mammalian Cells Cultures in vitro. Experimental Cell Research, S. 441-447.

Wayne L. Monsky, D. F. (15. August 1999). Augmentation of Transvascular Transport of Macromolecules and Nanoparticles in Tumors using Vascular Endothelial Growth Factor. CANCER RESEARCH, S. 4129–4135.

Wei, Z.-H., Lee, C.-P., & Lai, M.-F. (28. August 2009). Magnetic particle separation using controllable magnetic force switches. Journal of Magnetism and Magnetic Materials, S. 19-24.

Wen He, F. Y. (68 2012). Microbubbles with surface coated by superparamagnetic iron oxide nanoparticles. Materials Letters, S. 64-67.

Y. Zou, C. T.-S. (13 2007). p53 Aerosol Formulation with Low Toxicity and High Efficiency for Early Lung Cancer Treatment. Clinical Cancer Research, S. 4900 – 4908.

Yokoyama, M. (8 2005). Drug targeting with nano-sized carrier systems. Journal of Artificial Organs, S. 77-84.

Yolanda Sanchez-Antequera, O. M. (16. August 2010). Gene delivery to Jurkat T cells using non-viral vectors associated with magnetic nanoparticles. International Journal of Biomedical Nanoscience and Nanotechnology, S. 202-229.

Young-Eun Choi, J.-W. K. (6. January 2010). Nanotechnology for Early Cancer Detection. Sensors, S. 428-455.

Yuhong Mi, X. Z. (2006). Hydrothermal Fabrication of Spindle-type a-Fe2O3 Nanoparticle and its Magnetic Property. Proceedings of the 1st IEEE International Conference on Nano/Micro Engineered and Molecular Systems, (S. 519-522). Zhuhai, China.

Zijlstra, H. (1967). Experimental Methods in Magnetism. Amsterdam: North-Holland Publishing Company.

i want morebooks!

Buy your books fast and straightforward online - at one of the world's fastest growing online book stores! Environmentally sound due to Print-on-Demand technologies.

Buy your books online at
www.get-morebooks.com

Kaufen Sie Ihre Bücher schnell und unkompliziert online – auf einer der am schnellsten wachsenden Buchhandelsplattformen weltweit!
Dank Print-On-Demand umwelt- und ressourcenschonend produziert.

Bücher schneller online kaufen
www.morebooks.de

OmniScriptum Marketing DEU GmbH
Heinrich-Böcking-Str. 6-8
D - 66121 Saarbrücken
Telefax: +49 681 93 81 567-9

info@omniscriptum.de
www.omniscriptum.de

Printed by Books on Demand GmbH, Norderstedt / Germany